Mobile Media

Content and Services for Wireless Communications

Mobile Media

Content and Services for Wireless Communications

Edited by

Jo Groebel
Eli M. Noam
Valerie Feldmann

COLUMBIA INSTITUTE FOR TELE-INFORMATION

NEW YORK AND LONDON

Camera ready copy for this book was provided by the author.

First published by Lawrence Erlbaum Associates, Inc., Publishers

This edition published 2012 by Routledge

Routledge
Taylor & Francis Group
711 Third Avenue
New York, NY 10017

Routledge
Taylor & Francis Group
2 Park Square, Milton Park
Abingdon, Oxon, OX14 4RN

Cover design by Tomai Maridou

Library of Congress Cataloging-in-Publication Data

Transatlantic Dialogue (2nd : 2002 : Dusseldorf, Germany, and New York, N.Y.) Mobile media : content and services for wireless communications / edited by Jo Groebel, Eli M. Noam & Valerie Feldmann.

p. cm.

"Chapters of this book were initially presented at the Second Transatlantic Dialogue...held in Dusseldorf and New York City...hosted by the European Institute for the Media and the Columbia Center for Tele-Information"—Pref.
Includes bibliographical references and index.

ISBN 0-8058-4642-5 (cloth : alk. paper)
ISBN 0-8058-5880-6 (paper : alk. paper)

1. Telecommunication—Technological innovations—Congresses. 2. Wireless communication systems—Congresses. 3. Digital media—Congresses. 4. Digital communications—Congresses. I. Groebel, Jo. II. Noam, Eli M. III. Feldmann, Valerie. IV. European Institute for the Media. V. Title.

HE7604.T73 2002
384.5—dc22

2005050373
CIP

10 9 8 7 6 5 4 3 2 1

In memoriam
Axel Zerdick

Contents

Preface ix
Jo Groebel, Eli M. Noam and Valerie Feldmann

Introduction xi

Contributors xvii

Glossary of Terms xix

I. Technology and Infrastructure Models

1 How Can Anyone Afford Mobile Wireless Mass Media Content? 3
Timothy X. Brown

2 Always-On Demand—The Digital Future of Communication 21
Miriam Meckel

3 On the Myth of Convergence 33
Klaus Goldhammer

4 Automotive Telematics: Is it Time for a Renaissance or an Obituary? 45
Jonathan Lawrence

II. Content Models

5 Are There Content Models for the Wireless World? 57
Benjamin M. Compaine

6 Design Strategies for Future Wireless Content 69
John Kelly

7 Mobile News Design and Delivery 87
John V. Pavlik & Shawn McIntosh

8 Mobile Peer-to-Peer Content and Community Models 97
Valerie Feldmann

9 Contents and Services for Next Generation Wireless Networks 115
John Carey

III. Business Models
10 Profitable at any Speed? 133
Bertil Thorngren

11 Mobile Commerce Business Models and Network Formation 139
Carleen F. Maitland

12 Mobile Communications Business Model in the United States 153
James Alleman & Christopher Swann

13 Mobile Wireless Strategy of Media Firms: Examining the Wireless Diversification Patterns of Leading Global Media Conglomerates 165
Sylvia M. Chan-Olmsted & Byeng-Hee Chang

IV. Policy Models
14 Exclusive Rights in Information and Mobile Wireless Mass Media 187
Yochai Benkler

15 3G or Not 3G: The WiFi Walled Garden 201
Kenneth R. Carter

16 Emergency Communication Needs: Mobile Content 215
Jonathan Liebenau

17 Access of Content to Mobile Wireless: Opening the "Walled Airwave" 225
Eli M. Noam

Outlook
18 Mobile Mass Media: A New Age for Consumers, Business, and Society? 239
Jo Groebel

Index 253

Preface

Mobile wireless media today: clueless, speechless, and pennyless? Or on the verge of emerging as a new mass content medium? The answer is important economically and culturally. The concept of mobility has been powerful for voice telephony, where it is a worldwide phenomenon. But now, mobile communications aim to move beyond individualized voice and advance to a much more complex territory: mass media content—text, voice, sound, images, even video. A new medium may be emerging. If it manages to take off, it will create new types of content, generate new technologies, allow users to interact in new ways, develop new business opportunities and raise new issues of public policy. But it might also be a pipe dream, another instance of hot air fuelling another bubble. These are important issues to analyze, because the development of mobile media may substantially influence the media and wireless communications industry structure and industrial dynamics, as well as public policy.

Although a new medium may be born, there has been a dearth of academic analysis. Against this background, we have brought together an outstanding group of authors to contribute to our knowledge of the mobile Internet. The objectives of this volume, represented in its four sections, are:

- to analyze the emerging models for technology and infrastructure,
- to identify likely types of content such as news, entertainment, peer-to-peer, and location-specific information,
- to evaluate the economics, business models, and payment mechanisms necessary to support these media, and
- to cover policy dimensions such as copyright, competitive entry, and access by content providers.

Some underlying trends that the authors identify are the growing importance of the personalization of services, always-on features of mobile devices, emerging content models in unlicensed spectrum, and open access to content and to content sharing. However, the authors also cast doubt on the sustainability of a TV entertainment style mass media delivery model for mobile wireless, due to the opportunity cost of spectrum for individualized video use. For less spectrum-intensive usage, the ubiquity and customization potential of wireless media enable flexible forms of media that can fit personal needs and situations. Against this background the contributors discuss many facets of mass media content prospects for mobile wireless.

Most of the chapters of this book were initially presented at the Second Transatlantic Dialogue on 'Mass Media Content for Wireless Mobile Communications', held in Düsseldorf and New York City. European and American experts from industry, media, academia, and policy discussed the challenges of new media developments and their impact. The conferences were hosted by the European Institute for the Media and the Columbia Center for Tele-Information (CITI) of the Columbia Business School.

The European Institute for the Media is a think tank, which analyses media and communications development in Europe and advises policymakers, governments and other social groups on the constituents of the future "civil digital society".

The Columbia Institute for Tele-Information (CITI) is a university based research center focusing on strategy, management, and policy issues in telecommunications, computing, and electronic mass media. It is a Sloan Foundation industry research center.

Acknowledgments

Many people deserve recognition for contributing to the Second Transatlantic Dialogue and this book. We are grateful to our publisher Lawrence Erlbaum Associates, and its editorial director Emily Wilkinson, as well as to Nils Schleusner and Sergio Vitale from medionet AG, Berlin. We would also like to acknowledge our debt to Benjamin Bloom, Kenneth Carter, Gabriele Eigen, Darcy Gerbarg, Andrea Koenen, Bertram Konert, Rosa Morales, Nicole Reynolds, David Ward, and Stephanie Winde from EIM and CITI who were of enormous help. Finally we thank all the authors and contributors for their outstanding collaboration.

Düsseldorf and New York, January 2005

Jo Groebel
Eli M. Noam
Valerie Feldmann

Introduction

Mobile media are not a new phenomenon. Books, newspapers, and magazines; portable music players such as an MP3-player or portable game consoles; or just an ordinary car radio can be classified as mobile media. The relevant question is what is actually moving: information, devices, or people? The mobile media we focus on in this volume take advantage of the mobility of all three elements. People can move freely without being disconnected; devices are portable; and information moves freely and can reach specific recipients.

In this book we provide four different perspectives on the development of mobile media: technology, content, business, and policy. The authors of the various chapters provide innovative and often provocative thoughts on these four dimensions to the wireless world.

Technology and Infrastructure Models

Mobile media content services build upon technology models of a variety of wireless networks. The expected benefits of the migration from 2G to 3G cellular networks focus on the increase in bandwidth and in speed, the introduction of packet-switched always-on services, and location technologies. While numerous barriers slow this development down, a new breed of wireless technologies in unlicensed spectrum is emerging that creates new options.

Timothy X. Brown examines the cost of delivering multi-media content on-demand to end users via a wireless media. The analysis lays out the technological and spectrum constraints that will shape any wide-spread deployment of wireless multi-media applications. It considers only existing and near-term delivery technologies. Three representative contents are analyzed: real-time video, audio, and text news. The analysis suggests that even low-fidelity high-compression video and audio will remain expensive. However, while a brute force on-demand, anywhere, high-quality real-time content model may not be viable, alternative delivery models may prove quite economical.

Miriam Meckel analyzes "always on strategies" and "on demand strategies" against the background of the development of third generation (3G) mobile technologies in Europe. From the point of view of technology, basic trends include miniaturization, digitalization, networking and the further development of mobile communications structures. Closely related to these trends are social developments and changes in users' interests: globalization, mobility and flexibility, team work and knowledge transfer as well as changes in the relationship between the public and private spheres.

Klaus Goldhammer suggests that viewing developments of convergence in the wireless world is a myth. He identifies barriers to practicality and user-friendliness.

Jonathan Lawrence examines the market for mass media content and services through automotive telematics business models. The automobile is considered by many to be a major frontier for mass media content. Given that the time drivers spend in the vehicle is captive time, mass media companies in areas such as Internet content, satellite radio, and voice portals are searching for ways to capitalize on what they believe is a multi-billion dollar market opportunity. However, after several years and multiple independent and joint attempts by auto-makers, wireless carriers and information service providers, consumer adoption of automotive telematics remains well below forecasted levels.

Content Models

Even in mobile data growth markets such as Japan and South Korea, the introduction of new 3G services are only slowly gaining customers. Content as it is currently received via 2.5G technologies seems to offer enough value or in other words: content models for future wireless services still pose a number of questions even in the world's most advanced mobile wireless data markets.

Benjamin M. Compaine asks the key question "How does the mobility of wireless add value to content." The old wireless paradigm primarily revolves around the models of radio and television broadcasting. The new wireless paradigm has established itself around the point-to point model of telephony. But with huge sums being invested in licenses and infrastructures that can carry data as well as voice, service providers are expecting new revenue streams from content beyond telephony. Also, content providers—publishers and producers—are seeking new avenues for getting their substance out to end users.

John Kelly questions the familiar myth that "content is king". In terms of "social scope," he argues that successful wireless services will focus more on personal, or group, than mass media applications. He suggests viewing mobile wireless devices not as little, un-tethered TV sets, stereos, or computer screens, but as a new and different medium, with its own properties and potential. The best approach to designing new content and services is to envision a future "hypernet" in which mobile devices serve as a "remote control" and "intelligent agent" for life.

John V. Pavlik & Shawn McIntosh discuss mobile wireless technologies as a unique opportunity for media organizations, especially news and information providers, to create content customized for mobile receivers. In particular, the convergence of wireless, wearable, and locational technologies (e.g., the satellite-based global positioning system) make it possible to design and deliver location—or context-aware news and information to a mobile news audience. Such content would represent a significant transformation of the news and information paradigm that has dominated the market for mobile news design and delivery of the past century.

Valerie Feldmann suggests that mobile peer-to-peer communities will rather center around user-generated personal media files than professionally produced media content. However, mobile peer-to-peer communities have the potential to be used as promotional tools for media content, to strengthen the media brand identity and to contribute to users' brand loyalty. The value proposition for mass media content distributed via mobile peer-to-peer communities is classified along the dimensions of content availability, mobile community characteristics, and rules for sharing.

John Carey discusses demand uncertainties of the wireless landscape of content and services. One way to get a handle on the future is to explore the functionality of current applications, across cultures and across different types of user groups, and then to assess how core needs can be met in a new generation of wireless services. He argues that many desirable services are likely to be ones that enhance or build upon existing services such as messaging. The 'always on' feature of next generation wireless services may be as important as faster speeds.

Business Models

Paid content models are yet to be found for a wide range of Internet content services. Mobile data, however, already generates revenue via transactions, for example, logos and ringtones, subscriptions such as in the i-mode model, and data traffic charged by the wireless carriers. Yet, the volume of revenue is not sufficient to recoup investments in next generation networks and it is still unclear if it ever will. Business models that work in one national market cannot be easily translated into other markets. Evidence is given by the international expansions of the i-mode model that have yet to prove successful in different environments with regard to carrier strategies, consumer behavior, and regulatory frameworks.

Bertil Thorngren questions the reasoning that the demand for slow-speed mobile services can be transferred to high-speed mobile services. Increasing the speed and capacity in order to provide even richer content does not necessarily uncover new demand and new revenues to be shared among operators, content providers and other vendors. He asks how it is that the mobile operators have been so obsessed by the increase of sheer transmission speeds rather than other qualities where they can still claim a unique advantage. Other options like WLANs can offer radically higher speeds (Mbps rather Kbps) at a rapidly increasing number of Hot Spots, though nationwide coverage and roaming is still far off.

Carleen F. Maitland discusses the sources of firm power in mobile commerce networks and the implications of changes in this power distribution for mobile commerce business models. By examining the networks of industry players such as Vodafone, Vizzavi, NTT DoCoMo and Disney, industry trends in the power distribution among content providers and network operators are observed. Furthermore, the examination demonstrates that a variety of business models for mobile commerce are being pursued and that these models are subject to change.

James Alleman & Christopher Swann provide an overview of the business models for voice services of United States mobile communications providers. A mature demand for voice service is confronting the industry in the face of cutthroat pricing of services. He characterizes the business model that the wireless carriers have pursued as depending on lock-in of customers through subsidized handsets, low entry prices, bundled minutes, a lack of number portability and non-compatible handsets.

Sylvia M. Chan-Olmsted & Byeng-Hee Chang assess the product and international diversification strategies of the leading global media conglomerates in the mobile wireless market. These conglomerates have approached the mobile wireless medium with limited asset diversification but numerous strategic alliances to improve the wireless accessibility of their media content, brands and Internet services. The factor of product relatedness apparently presents an obstacle for aggressive extension to the wireless sector by the media conglomerates.

Policy Models

Mobile communications markets are widely considered to be competitive. However, as mobile communications' role in relation to fixed-line communication is getting stronger and voice and data traffic to and from mobile phones is increasing, there are a number of policy issues that need to be addressed. New approaches in spectrum policy, interconnection issues, competition policies for open access, and intellectual property rights are posing challenges for new regulatory policies.

Yochai Benkler writes on the implications of exclusive rights in information (ERIs) for the future of mass media content delivery to mobile wireless devices. He opposes the idea for mobile wireless data communication to replicate the mass media model. Instead, he suggests that the core of mobile wireless communications should be, and will be, the provision of high-speed mobile Internet access through equipment that will enable license-free, network-owner-independent communications, not mass media.

Kenneth R. Carter analyzes business strategies of communications carriers related to the system of spectrum allocation whereby carriers purchase the exclusive right to use spectrum and new entrants employ strategies in unlicensed spectrum. These unlicensed networks present a threat to the profitability and commercial viability of existing cellular networks and emerging 3G networks. Thus, he argues that next generation service providers will have to integrate licensed and unlicensed spectrum in their networks and find means of recreating the barriers to entry of licensing regime with differentiation, externalities, and network investment.

Jonathan Liebenau analyzes mobile content appropriate for emergency needs that could take a variety of imaginative forms. Yet, the feasibility, utility and the responsibility to provide and maintain such content raises questions of cost, practices of use, and requirements. He looks at static data such as building and neighborhood plans and other forms of geographical information systems data as well as dynamic data including data from sensors (heat, smoke, water, movement, radiation, etc.) and information passed to those in need by emergency workers, such as escape advice and situation reports.

Eli M. Noam discusses the issue of access, diversity, and openness of content providers. He argues that openness is more than competition of wireless service providers. Openness means the ability of content providers to access consumers without gatekeepers. This is always an issue in network industries, and has been a constant theme of regulatory battles for over a century. Yet, openness to customer equipment has not quite reached wireless communications. In the case of wireless, an end-to-end control of network operators creates potential bottlenecks for rival service providers, content, portals, and transactions.

Outlook

In the Outlook section, **Jo Groebel** describes his visions of mobile media's potential effects on consumers, businesses, and society. He suggests that intelligent mobile terminals will allow for local and flexible media use and will foster a development from *multimedia* to situation-dependent *polymedia.* Psychological dimensions of human behavior such as mobile mood management are in particular linked to this new situation-to-person (S2P) paradigm.

In the following chapters, we present innovative research on mobile media that both answers and opens up new questions in this exciting, and, for the media and telecommunication industries, relevant area.

Contributors

Editors and Authors

James Alleman, Professor University of Colorado, and Professor and Research Director, Columbia Institute for Tele-Information, Columbia Business School

Yochai Benkler, Associate Professor and Director, Engelberg Center for Innovation Law and Policy, New York University Law School

Timothy X. Brown, Assistant Professor, Department Interdisciplinary Telecommunications, Electrical and Computer Engineering, and Computer Science, University of Colorado

Benjamin M. Compaine, Research Consultant, Program on Internet and Telecoms Convergence Consortium, Massachusetts Institute of Technology

John Carey, Managing Director, Greystone Communications

Kenneth R. Carter, Deputy Director, Columbia Institute for Tele-Information, Columbia Business School

Sylvia M. Chan-Olmsted, Associate Professor, Department of Telecommunication, College of Journalism and Communications, University of Florida

Byeng-Hee Chang, Ph.D. Student, College of Journalism and Communications, University of Florida

Valerie Feldmann, Visiting Scholar, Freie Universität Berlin, and Affiliated Researcher, Columbia Institute for Tele-Information, Columbia Business School

Klaus Goldhammer, Managing Director, Gold Media

Jo Groebel, Director-General, European Institute for the Media, and Professor and Chairman, Department Mass Communication, Utrecht University

John Kelly, Principal Investigator, Interactive Design Labs, Columbia University

Andrea Koenen, Project Manager, Digital World Program, European Institute for the Media

Jonathan Lawrence, President, Crystal Mountain Group

Jonathan Liebenau, Senior Lecturer in Information Systems, London School of Economics, and Visiting Scholar, Columbia Institute for Tele-Information, Columbia Business School

Carleen F. Maitland, Faculty of Technology, Policy, and Management, Delft University of Technology

Shawn McIntosh, Adjunct Faculty, Department of Mass Communication, Iona College

Miriam Meckel, Professor and Permanent Undersecretary for Europe, International Affairs, and Media, Government Office of Northrhine-Westphalia

Eli M. Noam, Professor of Finance and Economics and Director, Columbia Institute for Tele-Information, Columbia Business School

John V. Pavlik, Professor and Executive Director, Center for New Media, Columbia School of Journalism

Christopher Swann, Senior Economist, Global Insight

Bertil Thorngren, Professor and Director, Center for Information and Communications, Stockholm School of Economics

Glossary Of Terms

2G	Second Generation Mobile Network or Service
2.5G	Second Generation Enhanced Mobile Network or Service
3G	Third Generation Mobile Network or Service
ABC	Always Best Connection
ARPU	Average Revenue Per User
ASP	Applications Service Provider
CDMA	Code Division Multiple Access
CBDTPA	Consumer Broadband and Digital Television Promotion Act
CMRS	Commercial Mobile Radio Service
COD	Cash On Delivery
DBS	Direct Broadcast Satellite
DMCA	Digital Millennium Copyright Act
DSL	Digital Subscriber Lines
DRM	Digital Rights Management
EMS	Enhanced Messaging Service
ERI	Exclusive Rights in Information
FCC	Federal Communication Commission
FOMA	Freedom of Multimedia Access
GCR model	Glorified Car Radio model
GPRS	General Packet Radio Service
GPS	Global Positioning System
GSM	Global System for Mobile Communications

HTML	Hypertext Markup Language
IEEE	Institute of Electrical and Electronics Engineers
IM	Instant Messaging
IP	Internet Protocol
IPR	Intellectual Property Rights
ITS	Intelligent Transportation Systems
J2ME	Java version 2, Micro Edition
LBS	Location-Based Services
LCR	Least Cost Routing
MHIA model	Mobile High-Speed Internet Access model
MJW	Mobile Journalist's Workstation
MMS	Multimedia Messaging Service
MOU	Minutes of Use
MPEG	Moving Pictures Experts Group
MVNO	Mobile Virtual Network Operator
OEM	Original Equipment Manufacturer
PCS	Personal Communication Services
PDA	Personal Digital Assistant
PSTN	Public Switched Telephone Network
RFID	Radio Frequency Identification
RPP	Receiving Party Pays
SDR	Software-Defined Radio
SIM	Subscriber Identity Module
SMS	Short Message Service
TDMA	Time Division Multiple Access
TSP	Telematics Service Provider
UMTS	Universal Mobile Telecommunications System
UWB	Ultra Wideband
WAP	Wireless Application Protocol
WiFi	Wireless Fidelity

WIP	World Internet Project
WISP	Wireless Internet Service Provider
WLAN	Wireless Local Area Network
WML	Wireless Markup Language
XML	eXtensible Markup Language

I

Technology and Infrastructure Models

1 How Can Anyone Afford Mobile Wireless Mass Media Content?

Timothy X. Brown

1 Overview

This chapter examines the cost of delivering multi-media content-on-demand to end users via a wireless media. The analysis lays out the technological and spectrum constraints that will shape any widespread deployment of wireless multi-media applications. It considers only existing and near-term delivery technologies. It does not address customer demand, the end-user terminal, nor content specific issues such as content generation, cost of content, billing models, and copyright. It takes as a baseline the current second generation wireless cellular architecture in the U.S. Three representative contents are analyzed: real-time video, audio, and text news. The analysis suggests that even low-fidelity high-compression video and audio will remain expensive at $0.12–$0.28 per minute. Higher fidelity would require a significant increase in bandwidth and would be five to 40 times more expensive. Mainly text news, on the other hand, requires little bandwidth and would be inexpensive.

Various technological advances, such as one-way only transmission, third generation wireless, and non-real-time delivery, could collectively reduce the cost by a factor of approximately 8. Significantly more savings are possible in a broadcast-like model that would share the delivery cost among many users. Alternatively, WLAN-based architectures that restrict access to Internet-like content downloads in localized hot-spots are an affordable delivery model. Therefore, while a brute force on-demand, anywhere, high-quality real-time content model may not be viable, alternative delivery models may prove quite economical.

The chapter begins by developing a simple technology-based economic model for delivering content via current cellular technology. It then examines alternatives for reducing the delivery cost.

2 The Economic Challenge

In this section we will establish that under current wireless deployments, individual on-demand high-quality audio or video delivered to wireless mobile users would be prohibitively expensive. Further we will establish baseline technology parameters which will clarify the driving factors behind the cost and be the starting point for our discussion of potential solutions to this problem in the next section.

Our strategy is to estimate the price charged by current wireless cellular providers to deliver a given bit rate, the so called *bit rate price*. We make a number of rather gross assumptions in this analysis, but, as we will see, even, if the price estimates are off by a factor of two or three, the conclusions would remain unchanged.

Wireless cellular in the U.S. is quite competitive with up to nine different service providers in every major market.[1] With this level of competition, we assume that profits are limited and the price approximates the cost of providing the service. The cost of providing the mobile telephone service, as we will establish in estimating the bit rate price, is, to a first order, proportional to the bit rate of the service. Therefore, by comparing the bit rates of different services to the bit rate price, we can estimate the price at which these services would be offered.

2.1 Estimating the Bit Rate Price

The price per minute of a mobile telephone call divided by the data bit rate of the telephone call yields a price per bit rate for the phone call. In this section we will estimate the price per minute and bit rate for voice, and then show that the resulting price per bit rate applies to other mobile communication services at other bit rates.

The price per minute of use (MOU) for a mobile telephone call could be estimated through the many different service pricing plans. Unfortunately, these plans do not correctly state the price per MOU. For instance, if someone pays $0.10 per MOU for 300 MOU and only uses 150 minutes, then their effective price per MOU is $0.20. Further, prices change between peak and off peak times. We do not have the data to find true prices and usage times for individuals. In particular, we do not have data limited to peak demand periods that would best reflect the marginal cost of providing a minute of service. However, we can still estimate a price per MOU through aggregate statistics. A survey of consumer data results in an FCC study (FCC, 2003) yields an average monthly MOU that has risen from 380 minutes in 2001 to 427 minutes in 2003. The average subscriber monthly bill has risen from $47.37 in 2001 to $48.40 in 2002. The total bill divided by the monthly MOU is an upper bound on the marginal price of one MOU since the monthly bill includes fixed cost such as billing. If instead we look at the yearly increase in MOU and the yearly increase in the monthly bill we can derive a lower bound estimate on the marginal price of an MOU. This is a lower bound since price per MOU is falling, and tends to depress the overall monthly bill. From the above data, we derive the following estimate on the price per MOU:

$$\$0.11 = \frac{\$48.40}{427\text{MOU}} > \text{PriceperMOU} > \frac{\$48.40 - \$47.37}{(427 - 380)\text{MOU}} = \$0.022 \qquad (1)$$

Next we consider the bit rate of a mobile telephone call. In the U.S., most mobile phones are digital and based on one of three air interface technologies, the so-called USDC (IS-54), CDMA (IS-95), or GSM. Other digital technologies, such as PACS and iDen represent less than 10% of the total market and so are not considered. Next generation standards will be discussed later.

Table 1 shows the raw channel bit rate, the raw per user bit rate after removing control, synchronization, and signaling overhead, and the final usable data rate after removing error correcting coding overhead.[2] We don't consider any additional overhead specific to multimedia delivery (Radha et al., 2000).

Table 1: Various bit rates of 2G standards (Rappaport, 2002)

standard	raw channel bit rate (kbps)	per user bit rate (kbps)	usable user bit rate (kbps)
GSM	270.8	22.8	13.0
USDC	48.6	13.0	7.95
CDMA	1228.8	14.4	6.65

The only relevant rate here is the usable user bit rate (the others are included to suggest the overhead that is necessary to enable a high-quality wide-area high-mobility radio connection). This shows a range of 6.65kbps to 13kbps for these different standards. Combining these values with (1) we come up with the following bound.

$$\$0.017/\text{kbps} = \frac{\$0.11}{6.65\text{kpbs}} > \text{price per bit rate} > \frac{\$0.022}{13.0\text{kbps}} = \$0.0017/\text{kbps}$$

We are only interested in a gross estimate so we simplify these bounds to:

$$\text{Price per minute per bit rate} = P_b = \$0.01/(\text{min kbps}) = \$10.00/(\text{min Mbps}) \quad (2)$$

The result in (2) can be interpreted to say that for typical cellular and PCS deployments in the U.S., the cost to deliver a reliable 10kbps data rate to a user is \$0.10 per minute. It should be realized that this cost is a cost per volume of data, that is:

$$\frac{\$0.10}{\text{min}\,10\text{kbps}} \times \frac{1\,\text{min}}{60\,\text{sec}} \times \frac{8\text{bit}}{\text{byte}} \times \frac{100\text{kB}}{\text{MB}} = \$1.33/\text{MB} \quad (3)$$

In some cases, such as pricing news delivery (i.e., downloading a known size file), we will work directly with this volume price, and not concern ourselves with the delivery speed.

2.2 Estimating the Service Price

The result in (2) can be further extended to show that it scales over a range of bit rates, that is, $P(B)$, the price to offer a service at a given bit rate, B, is simply:

$$P(B) = P_b\, B. \quad (4)$$

By estimating the bit rate of different services, we can estimate the price of offering these services.

In so-called 2.5G standards (such as the IS-54 variant, IS-136, or the GSM variant GPRS) users can receive different data rates. IS-136 is a time division multiple access (TDMA) scheme with 6 time slots per frame. A normal user receives two time slots per frame to yield a 7.95kbps usable data rate. A user who wanted half the rate would use only one time slot, and a user who wanted 3 times the data rate would use all 6 slots. GPRS which is also TDMA behaves similarly, i.e. a user who doubles their data rate doubles the resources taken from the network (Rappaport, 2002). In IS-95, users can use greater or lower bit rates, and as in the TDMA system, a user consumes communication resources in proportion to the bit rate (Gilhousen et al., 1991).

If users double their bit rate they would require twice the capacity resources. As argued earlier, mobile telephone service has a high degree of competition in the U.S.. Further, capacity in terms of base stations is increasing at 20% per year (Roche et al., 2004) in order to keep up with demand. There is no reserve of excess capacity to accommodate higher data rate customers. Therefore, in order to cover their costs, service providers would need to double their price to users with twice the bit rate as indicated in (4).

This argument does not extend indefinitely. For instance, very low data rate services like short messaging services use a different packet switched-like model for delivery. At very high data rates we exceed the maximum bit rate per channel of the standard. For our purposes though, we assume that the price scales linearly with required bit rate as needed. For instance, very high data rates could be accommodated by using multiple channels.[3]

Based on this notion, what are the costs of providing our baseline real-time services of high-quality stereo audio and high-quality full-frame video? The bit rate of near CD-quality MPEG layer 3 (MP3) audio is 128kbps (Puri et al., 2000a). Similarly for television quality MPEG 1 video with stereo sound, the data rate is 1.2Mbps. For news, the New York Times home page is approximately 150kB.[4] Applying these to (3) and (4), yields $1.28/min for audio, $12.00/min for video, and $0.20/download for news.

What do these prices mean? Watching a 100 minute movie would cost $120. Listening to a single three-minute song would cost $3.84. Downloading 10 news stories in the process of reading the morning news would cost $2.00. This is above and beyond the price of the content. It is unlikely that content providers would agree to a revenue split that is so heavily weighted toward content delivery. So, it is unlikely that content would be available even if some segment of consumers were willing to pay these prices.

Clearly these prices must be significantly reduced to achieve mass market acceptance. So, what can be done? The basic problem is that the bit rate demand is high compared to current providers capacity. So broadly, we can consider two approaches: lowering the user bit rate demand or lowering the price by increasing capacity resources. In other words, in (4) we can reduce B or we can reduce P_b. The next two sections discuss each of these in turn.

3 Reducing the Bit Rate

This section describes several methods of reducing the effective bit rate from which the reduction in the price of the service can be calculated directly . The methods considered include lower bit rate encoding, one-way vs. two-way communication, and better channel spectrum efficiency.

Lower Bit Rate Encoding: It is difficult to reduce the bit rate through encoding while preserving the content quality. Current encoding of audio and video content as well as the images that make up the bulk of web page downloads is relatively compact.

Better high-quality coders are an active area of current research. For instance, the H.26L standard is being developed to work at 800kbps and research shows that rates including high quality stereo audio could be pushed to 400kbps (Dumitras & Haskill, 2002). The lowest of such rates are computationally intensive and can not be encoded in real time. This suggests that we might expect a factor of two reduction in the bit rate in the near term.

One-way vs. Two-way: Another way to reduce the effective bit rate is to use the asymmetry in the communication. The price per bit rate has assumed a symmetric two-way up and down link connection. But audio and video content is typically unidirectional and therefore uses one half of the spectrum. This can reduce the service price by half.

One difficulty in implementing such a strategy is that the current mobile cellular requires a two-way connection for connection maintenance. Mobiles assist in handoffs, acknowledge system commands, and so on. This overhead is necessary for connection quality and could not be completely removed; reducing the possible savings.

Better Spectrum Efficiency: Another approach is to use different radio techniques which in effect increase the number of mobile telephone channels per base station. For instance, if a new radio technology were deployed that could double the number of channels per base station at no extra cost, then we could halve the cost per channel. Since much of the cost is bound up in the base station, spectrum, and backhaul, the cost of a new radio technology would be a minority factor in the overall cost.

Unfortunately, we do not foresee any radio technology with significant spectrum efficiency improvements in the near term. The most prominent technology development is so-called third generation wireless (3G). CDMA2000 is the next generation of IS-95. Currently CSDMA2000 has two versions being deployed, CDMA2000 1X which is similar to IS-95, but allows shared user rates up to 144kbps; and CDMA2000 EV-DO which has shared user rates up to 2,400kbps. Early tests suggest effective rates for the EV-DO version of 400–800kbps (Seybold, 2002). Another 3G technology, WCDMA, is promoted by GSM carriers. It has rates similar to CDMA 2000 EV-DO (although using 3 times the spectrum). WCDMA is not expected to have widespread deployment for five years (Seybold, 2002). Third generation wireless technologies are primarily focused on providing a wide variety of data rates and services and are projected to have only a factor of two improvement in spectrum efficiency (Nettleton, 2002; Seybold, 2002). Recent pricing announcements suggest that prices for newer technology will be close to (3). For instance, AT&T Wireless prices 1MB of data at $1 for high volume users (ATT, 2004).

Other technologies on the horizon such as so-called 4G and multibeam antennas have promise to provide large efficiency gains. But they are considered to be still 5 to 10 years out and so outside the scope of this chapter (Gitlin, 2002). Satellites provide mobile service. Generally, service prices are higher than cellular. Satellites are not a solution to the economic challenge.

Combining all the near-term advances suggested in this section, we might expect a factor of 8 reduction in the cost of delivering the multimedia content. These advances suggest a cost of $0.16/min for audio, $1.50/min for video, and $0.025 per page of news.

4 Increasing Capacity

The bit rate price is affected by limits in spectrum and capacity for providing service. If the capacity could be increased cost-effectively, the bit rate price may be able to be reduced. This section looks at ways to increase the effective capacity. It analyzes the availability of more licensed spectrum, the potential for using unlicensed spectrum, and the cost of more base stations.

4.1 More Licensed Spectrum

Currently the U.S. has 180MHz of spectrum for mobile telephony; 50MHz in the 800MHz cellular band, 120MHz in the 1900MHz PCS band, and at least 10MHz in the 800MHz ESMR band. This section explores other bands that have potential for offering mobile communication services both now and in the future.

The first constraint is allowed frequencies. Frequencies above approximately 3000MHz are unsuitable for a mobile wireless application because they require line-of-site or near line-of-site to maintain a connection. Low frequencies below approximately 100MHz are not suitable since efficient antenna sizes become unwieldy for a low power mobile device.

Between 100 and 3000MHz other licensed bands include the MDS bands at various frequencies between 2.150 and 2.680 GHz (FCC, 2001a), and the WCS bands at 2.3GHz (FCC, 2001b). While these bands constitute a large amount of spectrum (108MHz), they are limited to fixed and not mobile applications. The significance of these limitations is reflected in the bidding which resulted in only $200M total from operators.

The FCC is licensing more spectrum for mobile applications via spectrum auctions. Potential spectrum here includes the so-called lower and upper 700MHz Bands. These have 78MHz of bandwidth between them and are located in an ideal band with respect to propagation and radio equipment. They are currently encumbered with existing TV stations especially in urban markets, but these stations are expected to migrate to lower frequency channels by 2006 (FCC, 1997).

The FCC and NTIA have proposed sharing plans for both the MDS bands and other bands currently controlled by the federal government. Under the rules, 3G-like services could be offered which meet certain interference limitations with the current incumbents. These limitations are so onerous that it is not believed that any major metropolitan market would be able to offer 3G service with this band (Weingardt & Murphy, 2001).

The recent C and F block broadband PCS auctions generated $16.9B for up to 30MHz of spectrum in each market[5] (FCC, 2001d). The bulk of the revenue was generated in the densest urban markets. For instance, the 30MHz of licenses for New York City generated $5.6B ($300/pop)[6] and similarly Los Angeles generated $1.5B ($200/pop). The 30MHz in these two auctions represents a 20% increase (150MHz to 180MHz) in available spectrum. These bids suggest significant increases in mobile wireless spectrum must be made before spectrum demand becomes saturated. This suggests that new licensed spectrum is an expensive way to add spectrum, on the order of building more base station infrastructure as described later.

4.2 Unlicensed Spectrum

Unlicensed spectrum provides an opportunity for expanding the available spectrum without the costs associated with the spectrum auctions. In the U.S., there are three prime unlicensed bands in the 100 to 3000MHz range with significant bandwidth: The 902 to 928MHz ISM band, 1.91 to 1.93GHz PCS bands, and the 2.4 to 2.4835GHz ISM band (FCC, 2001d).

The 900MHz band is used by cordless phones and was used by Metricom to offer their Ricochet service. Power limits for the band limit the effective range. The PCS unlicensed band is the smallest band and has used restrictions that would limit widespread use of the band.

The 2.4GHz is a large band which has proved quite successful for wireless local area networks (WLAN). Some operators are extending this model to provide fixed wireless Internet access. Wide-area Internet access is threatened in the long term as increasing interference from more users reduces the ability to make long connections. The more promising application is for providing hot spot access at hotels, airports, restaurants, etc. This relies on short range connectivity using a plethora of low-cost 802.11 access points. We will expand on this idea later.

4.3 More Base Stations

The wireless cellular concept allows operators to increase capacity within a fixed amount of spectrum by building more and more base stations (Rappaport, 2002). For a given amount of spectrum, each base station has a fixed quantity of capacity. The capacity of an operator's system is proportional to the number of base stations.

Table 2 shows the U.S. mobile telephone industry's annual capital investment, increase in number of base stations, increase in number of subscribers, investment per added base station, and investment per net added subscriber for the past 10 years. The total investment per base station has generally held close to $1M per base station for a decade. A look at investment per subscriber tells a similar story where investment per added subscriber has fluctuated around $900. Over this period, the cost of adding new capacity has been constant (to within a factor of 2).

Table 2: Mobile Telephone Industry Annual Added Cell Site and Subscriber Costs for the U.S. 1991–2001 (Roche et al., 2004)

year	investment (millions)	increase in cell sites	increase in subscribers ('000)	investment per cell site	investment per added subscriber
1994	$4,983	5,096	8,125	$978,000	$613
1995	$5,141	4,743	9,652	$1,084,000	$532
1996	$8,494	7,382	10,260	$1,151,000	$828
1997	$13,480	21,560	11,270	$625,000	$1,196
1998	$14,480	14,290	13,900	$1,013,000	$1,042
1999	$10,720	15,810	16,840	$678,000	$637
2000	$18,360	22,590	23,430	$813,000	$784
2001	$15,410	23,250	18,896	$663,000	$815
2002	$21,900	11,800	12,392	$1,856,000	$1,767
2003	$18,940	23,650	17,955	$801,000	$1,055

The $37B cost of spectrum licenses adds a fixed cost shared by every base station and every customer. June 2004 had an estimated 148M subscribers and 150,000 cell sites. These numbers yield an added $250/subscriber and $250,000 per cell site. Both of these numbers are smaller than the capital outlay number and will decrease with continued growth. While significant, the added spectrum cost does not change the conclusion that the cost per cell site or subscriber will not decrease significantly in the near future.

In this section, we have shown that adding spectrum or base stations are expensive propositions that will not likely reduce the bit-rate price.

5 Quality

So far we have considered the cost of delivering on-demand, anywhere, high-quality real-time content. This section and the next three sections consider the gains from relaxing these requirements along each of the three dimensions: quality, choice, location. First we consider the role of high-quality content.

Multimedia content can trade off quality for lower bit rates in a very direct way. Audio can be reduced to very low rates. Mobile telephony encoding speech (but not general audio such as music) at approximately 10kbps has been the starting point of this analysis. General audio rates between 12 and 256kbps can be achieved where 12kbps is for near telephone quality mono and 265kbps is for high fidelity stereo (Audioactive, 2002; Bouvigne, 1998).

Lower quality video encoders reduce the bandwidth by reducing the screen resolution, reducing the number of frames per second, or allowing more encoding artifacts to be introduced into the screen. For instance, video conference quality is 384kbps and a graing thumbnail is 28kbps (Bhaskaran & Konstantinides, 1997; Puri et al., 2000b).

For news, the main content is text. But much of news web pages are made up of ads and figures. The text content of the New York Times home page and top news stories consists of approximately 6kB of text per download.

The lowest bit rates are 28kbps for video, 12kbps for audio, and 6kB/download for news. Combining these rates with the near-term advances in reducing the bit rate price (the factor of two for one-way transmission, and the factor of two for better spectrum efficiency), these translate into $0.07/min, $0.03/min, and $0.002/download.

The results in this section show text news can be quite cost effective at a fraction of a cent per download. Audio and video can be brought to below current prices per minute for cellular voice. But, the quality of the telephone-grade audio and thumbnail video is so low that it is not likely consumers would be willing to pay anything. The higher quality content has 4 to 40 times higher bit rates, and would have as a result prices which are again too expensive.

6 Location

The delivery model has so far assumed that consumers could download content anywhere they can make a mobile phone call; inside or outside, urban or rural areas, standing still or moving at high-speeds. In this section we consider a model suitable only for stationary users at localized coverage locations.

WLAN's are being deployed at many levels by end users, private service providers, free service providers, and paid service providers. We will focus on the WLAN standard which is by far the most popular, 802.11b also known as WiFi.

WiFi networks are spreading rapidly. Over 40% of large corporations were using wireless networks at year end 2002. End users are buying WiFi interface cards in great numbers. Laptop computers come with integrated WiFi interfaces. Further, many are installing WiFi networks within their homes (Werbach, 2001).

Private service providers include campuses and offices which offer the service to a limited proscribed community such as students or employees. Many organizations, especially in the high-tech sector, are deploying such networks in this manner.

Free service providers include restaurants, individuals, and some organizations that choose not to restrict access or charge for providing connectivity. Many public spaces have service provided in this way. Perhaps the most interesting phenomenon here is the emergence of "community networks" consisting of individuals who make their wireline access available to all.

Finally, paid service providers sell WiFi-based access to subscribers. Often these services are deployed in airports, business hotels, and the like, catering to business travelers.

The main point here to note is that operators do not pay for their spectrum, equipment is cheap, and the marginal impact of additional users on low cost DSL and cable modems is negligible. As a comparison Verizon Wireless offers broadband access using 3G technology for $80 per month. T-Mobile offers a similar service at WiFi hotspots for $30 per month.

The unlicensed spectrum is subject to unregulated interference but, if intended coverage is kept localized around the access points, radio link budgets can retain a sufficient interference margin. In other words, the opportunity cost to existing WLAN operators for a new WLAN access point is negligible if the coverage of the new access point is localized. The net effect is that each additional access point adds more capacity similar to adding base stations in cellular.

WLAN cards are under $100 and access points are less than $300.[7] Access point installation often consists of attachment to a wall or ceiling and running a cable to the network interface. Some vendors sell integrated wireless access points and DSL/cable modems. The cost of high-speed access for heavy business users is $300/month.

To put these costs in perspective, suppose that we want nationwide coverage which we define as one access point for every 20 persons in the U.S. This yields approximately 13 million access points. Conservatively, we estimate a $1000 cost for each access point and installation and each access point requires a single high-speed access connection. This yields a $13B installation cost and an operating cost of $47B per year. The installation costs are comparable to the annual wireless capital investment listed in Table 2. The $47B operating cost is comparable to the $88B in wireless cellular revenue from in 2003 (Roche et al., 2004). Unlike the concentrated investment in Table 2, these costs are distributed across many individuals and organizations as suggested above.

To look at it another way, suppose each access point is used during one busy hour each day. Under the above assumptions, the cost of the access point and one year high-speed data service is $4600. If each access point is used for one busy hour per day, this is 22,000 minutes of use per year. Assume that the throughput of the access point and high-speed modem is 1Mbps (conservative given the 11Mbps peak rate). If the $4600 is to be recovered in one year, then the cost per minute of use per bit rate is $0.21/(min Mbps). Comparing this with the result in (2), this suggests a bit rate price that is 50 times lower.

Though significantly cheaper, the service provided is not the same as in the cellular model. The coverage is localized to the hot spots and not wide area. The users must be stationary or have limited mobility during the content delivery. The connection service is more similar to today's internet. In particular, no quality of service guarantees are provided like in the circuit switched (i.e., reserved bandwidth) model of traditional cellular. As in streaming media over the Internet, significant buffering is necessary to account for the more erratic internet performance. Though not a direct substitute for cellular delivery, it may have an important role.

7 Choice

There is in fact a very efficient mechanism for delivering wireless multi-media content: broadcast radio and television. Here the marginal delivery cost per user is zero. In the broadcast extreme, the delivery is efficient, but the content choice is limited in both what can be viewed and when it can be viewed.

7.1 Content Sharing Models

Broadcasting is efficient since every user shares a single transmission. Content on demand on the other hand requires one transmission per user. We will discuss several models intermediate to these alternatives.

One method to increase time flexibility is to broadcast time shifted versions of the same content. For instance, a 90 minute movie could be broadcast on 9 channels each shifted by 10 minutes. A user would never have to wait more than 10 minutes to watch the movie. Further, they could fast forward, rewind, and pause in 10 minute increments. As long as at least 9 users shared the content, it would be a net win in spectrum efficiency.

Another approach is to aggregate content on demand requests. For instance, if two users ask to watch the same movie, then a single stream would serve both. This model could be facilitated by synchronizing starting points. For instance, all movies would start on the hour. As the hour approached, users would submit their requests, and then only one transmission would be required for each distinct request.

A third approach would be to offer many different content alternatives but only broadcast the content actually being viewed. For instance, users would be offered a choice of 1000 different channels, but, in a single cell, if only 10 different channels are actually being viewed, only those 10 are broadcast. More detailed user preference and traffic data would be necessary to show the viability of such a model.

The main principle here is that the cost per user decreases as more users share the content. In particular, the marginal delivery cost is zero after the first user. This suggests that pricing should be designed to encourage content sharing. Potentially popular content should be given a lower price in order to steer users to common content. Less popular content that is less likely to be viewed by multiple users should be more expensive to reflect its true marginal cost.

7.2 Real Time vs. Non-Real Time

In an immediate content-on-demand model subscribers can choose any content and receive it immediately. In the content sharing models, user choice and flexibility are reduced to encourage coincident requests. Alternatively, subscribers could choose a future playback time and the content could download over time as bandwidth is available. This approach would require significant storage and would not be appropriate for all user terminals. Non-real-time delivery can stop and resume during congestion times while real-time delivery requires a continuous commitment from the network once the streaming begins. Real-time content requests must be rejected if bandwidth is not immediately available.

The advantage of non-real-time delivery can be demonstrated with the following model. Requests for downloads arrive over time. The service provider has C channels set aside for the multimedia downloads. If the service provider has C or fewer requests, they are all streamed out. If more than C are active the excess are queued and served in order as active streams finish. Under standard

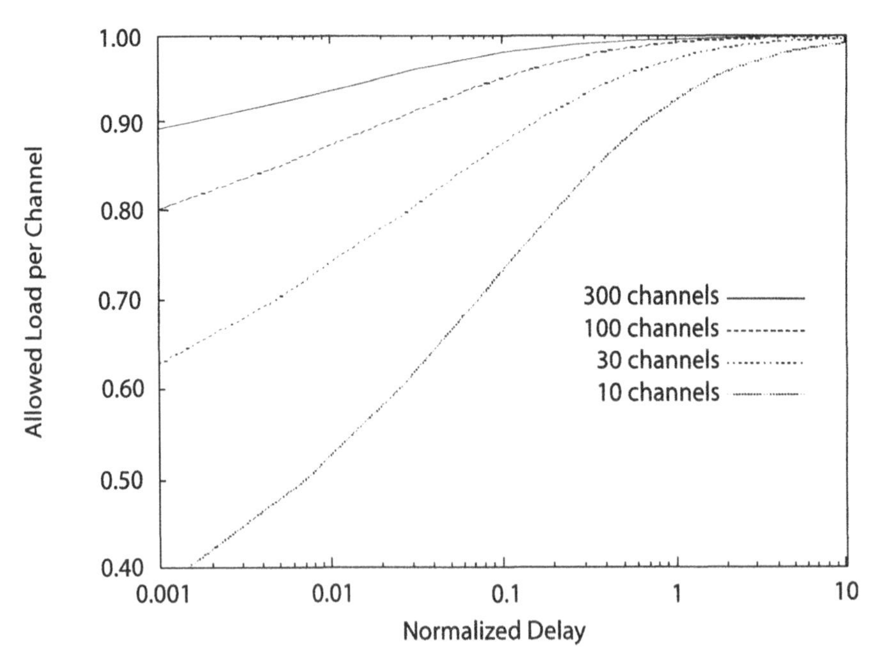

Figure 1: The normalized channel utilisation as a function of the wall time to service (normalized by the average time) for a non-real-time delivery model

traffic assumptions we can compute how much load the system can carry as a function of the average wait time before a request starts service.[8]

Figure 1 plots the channel utilization as a function of delay. For example, let resources for $C = 10$ video sessions be set aside and video sessions consist of 10 minute clips. If clips must be served within 6 sec, normalized average delay of 0.01, then the load can only be about one half. On the other hand, if users can wait 10 minutes (normalized delay of 1) then the load exceeds 90%.

This model shows that with a small number of available channels, utilization approximately doubles. In other words, non-real-time delivery is twice as efficient as real-time delivery. The efficiency increase can be even higher if combined with the content sharing models where common requests are aggregated and sent out as a single stream.

Perhaps more important than the increased efficiency, non-real-time delivery is more suitable for the WLAN model described previously which on average has large low-cost bandwidth, but over time can have significant performance variations which make it unsuitable for real-time delivery. The main drawback to non-real-time delivery is the lack of on-demand downloads and that the mobile device needs sufficient storage for the content.

8 So What Can Be Done?

Unmodified video and audio content on demand will never be economically viable over traditional cellular and PCS networks. Adding more costly spectrum or base stations will not change these economics. Their viability increases dramatically with compression and trading off lower fidelity for lower bandwidth. However, users will be less willing to pay for lower fidelity and it is not clear that a viable price/fidelity combination exists. Greater efficiency, such as with newer cellular protocols, improves this situation, but is not in itself enough. Many current wireless providers have taken this approach; offering low resolution images and video delivery.

For high-fidelity content what are needed are newer models for downloading. Broadly, two approaches appear viable. The first is to move from high-cost cellular to a low-cost WLAN based hot-spot approach to wireless access. The WLAN approach is better suited to a download now and view later model and can be very cost efficient. The other approach is closer to a broadcast model where the content that can be delivered is constrained to encourage content sharing. Both of these approaches remove much of the spontaneity of a content on demand model but appear to be the most promising approach to cost effective multimedia content delivery.

Endnotes

[1] Between the Cellular (2), PCS (6), and ESMR (1) bands any given area in the U.S. has as many as 9 different mobile telephone licenses Year end of 2000 data indicates that 83% of the U.S. population had a choice of 5 or more providers (FCC, 2003).

[2] IS-95 first adds signaling overhead and then adds coding overhead. Further, IS-95 uses a variable rate vocoder. We assume the vocoder works at 50% load on average.

[3] One of the 3G standards, CDMA-2000, uses a multi carrier technology that combines multiple channels for higher-rate users.

[4] Estimated from the size of the www.nytimes.com web-page downloaded on 03/13/02 and 03/14/02.

[5] These auctions offered licenses that were unsold or unpaid for in an earlier auction. So coverage was sporadic.

[6] The license price divided by the population in the licensed area.

[7] The prices in this section are based on a casual survey of various vendors and service providers at the time of writing. They are intended to be suggestive of the costs of WiFi service relative to conventional wide area cellular.

[8] In particular, requests arrive as a Poisson process, the average time to serve a request is exponentially distributed, and as a result the Erlang C distribution applies.

References

AT&T. (2004). Mmode plans. Retrieved September 6, 2004 from the World Wide Web: http://www.attwireless.com/personal/features/mmode/plans.jhtml.

Audioactive. (2002). *Audioactive internet bit rate versus quality (technote).* Retrieved October 3, 2002 from the World Wide Web: http://www.audioactive.com/intro/papers/bitrate.html.

Bassuener, K. (2002). Verizon wireless rolls out new data prices. *Wireless Week.*

Bhaskaran, V. & Konstantinides, K. (1997). *Image and Video Compression Standards: Algorithms and Architectures.* Kluwer, second edition.

Bouvigne, G. (1998). *Listening tests of mp3 using different bitrates.* Retrieved October 3, 2002 from the World Wide Web: http://www.mp3-tech.org/tests/gb/.

Dumitras, A. & Haskill, B. (2002). An encoder-only texture replacement method for effective compression of entertainment movie sequences. In *IEEE Internationsal Conference on Acoustics, Speech, and Signal Processing.* Orlando, Florida.

FCC. (1997). *Fifth report and order: Advanced television systems and their impact upon the existing television broadcast service.* Report and Order 97–116, Federal Communications Commission released April 27.

FCC. (2001a). *Auction 06: Multipoint/multichannel distributions services fact sheet.* Retrieved from the World Wide Web: http://wireless.fcc.gov/auctions/06/, updated August 2.

FCC. (2001b). *Auction 14: Wireless communication service (WCS) fact sheet.* Retrieved from the World Wide Web: http://wireless.fcc.gov/auctions/14/, updated August 2.

FCC. (2001c). *Auction 35: C and F block broadband PCS fact sheet.* Retrieved from the World Wide Web: http://wireless.fcc.gov/auctions/35/, updated September 12.

FCC. (2001d). *Unlicensed radio frequency devices.* Code of Federal Regulations Title 47, Chapter 1, Part 15, Federal Communications Commission. Revised October 1.

FCC. (2003). *Eighth annual report and analysis of competitive market conditions with respect to commercial mobile services.* Report 03-150, Federal Communications Commission. released July 14.

Gilhousen, K., Jacobs, I., Padovani, R., Viterbi, A., Weaver Jr., L., & Wheatley III, C. (1991). On the capacity of a cellular CDMA system. *IEEE Transactions on Vehicular Technology,* 40(2):202–211.

Gitlin, R.D. (2002). Challenges and issues for 4G/5G wireless networks. In *IEEE Wireless Communications and Networking Conference.* Orlando FL, March 17–21.

Nettleton, R. (2002). Third-generation cellular: Technical and economic challenges. In *2002 International Symposium on Advanced Radio Technologies.* Boulder, CO, 4–6 March.

Puri, A., Schmidt, R.L., & Haskell, B.G. (2000a). Overview of the MPEG standards. In Puri, A. & Chen, T. (Eds.), *Multimedia Systems, Standards, and Networks,* Chapter 4, (pp. 87–129). Marcel Dekker Pub.

Puri, A., Schmidt, R.L., Luthra, A., Chen, X., & Talluri, R. (2000b). MPEG-4 natural video coding—part I. In Puri, A. and Chen, T. (Eds.), *Multimedia Systems, Standards, and Networks,* Chapter 8 (pp. 205–244). Marcel Dekker Pub.

Radha, H., Ngo, C.Y., Sato, T., & Balakrishnan, M. (2000). Multimedia over wireless. In Puri, A. and Chen, T. (Eds.), *Multimedia Systems, Standards, and Networks,* Chapter 19 (pp. 525–557). Marcel Dekker Pub.

Rappaport, T. (2002). *Wireless Communications Principles and Practice.* Prentice-Hall, second edition.

Roche, R., Jobanputra, P., & Rodriguez, L. (2004). *CTIA's wireless industry indices, semi-annual data survey results, January 1985–December 2003.* Survey summary, Cellular Telephone Industry Association.

Seybold, A. (2002). Wireless Data University Training Seminar, Andrew Seybold Group, LLC. Las Vegas, October 15.

Weingardt, B.H. & Murphy, R. (2001). The search for 3G spectrum: Part II. *Wireless Broadband,* 2(6). June/July.

Werbach, K. (2001). The paradise of the commons. *Release 1.0,* 19(10).

2
Always-On Demand—The Digital Future of Communication[1]

Miriam Meckel

1 Introduction

The development of third generation (3G) mobile technologies remains a sensitive issue in Europe. The conflict between slower growth in the mobile communications markets and the high levels of investment needed to build UMTS networks has left the telecommunications industry facing enormous challenges. At the present time, it is not clear whether and when investments for UMTS licenses and for building the network will be recovered.

The origins of the public debate surrounding UMTS show that it is impossible to make accurate predictions regarding the digital future of communication. Short-term developments are usually overestimated while long-term effects tend to be underestimated. However, the developments we can observe in present-day technology and society provide some insight into the future of communication and highlight implications for future business and society models.

From the point of view of technology, basic trends include miniaturization, digitalization, networking, and the further development of mobile communications structures. Closely related to these trends are social developments and changes in users' interests: globalization, mobility and flexibility, teamwork, and knowledge transfer as well as changes in the relationship between the public and private spheres.

New, more diverse, and rapid means of communication are appearing. In a globalized world where great emphasis is placed on the mobility and flexibility of the individual, "always on strategies" ensure availability any time, any place. "On demand strategies" support this flexibility in terms of space and time.

Always-on demand? There is still some way to go before mobile platforms become widely accessible and the man/machine interface becomes easier to manage. Besides the issue of whether a technology proves to be useful in prac-

---

[1] In collaboration with Andrea Koenen.

tice, more important still is the social and cultural context of any new technology. Therefore the key to realizing future visions lies in the ability to combine technical capability with what is socially desirable.

2 Mobile Technologies: Forecasts Ranging from Euphoric to Gloomy

"The future of UMTS—From euphoria to disillusionment to financial burden" (Friedrich-Ebert-Stiftung, 2002); the title of this publication neatly sums up the way the public debate on 3G mobile technologies has developed. In August 2000, when German UMTS licenses were sold for almost € 50 billion, the marketplace was in a state of euphoria, eagerly anticipating a new era of communications and future profits.

Shortly thereafter the financial markets' confidence in the telecommunications sector began to wane. After a decade in which mobile telephony had determined the development of telecommunications markets, 2001 proved to be a turning point, raising many issues regarding the future of 3G (Knape, 2003). None of the major mobile communications groups (IDATE, 2002, p. 50) achieved their targets in 2001 and T-Mobile, KPN Mobile and BT Wireless had to postpone planned stock flotations (cf. Booz Allen Hamilton & GCI Hering Schuppener, 2001, p. 6). Insiders were skeptical about the future of mobile communications and were even asking themselves whether UMTS was "superfluous" (Friedrich, 2001).

Reasons given for the disillusionment include a lack of customer focus, the fact that new technologies and end products are not yet ready for market, the increasing complexity of integrated software and resulting susceptibility to error as well as the fact that a fundamental rule of business has been ignored: "Instead of identifying need, finding a solution, building up the business and then regulating it, with 3G attempts have been made first to regulate it, then to create a business and find a solution, while all the time trying to identify customer need." (Hürlimann, 2001).

Therefore, mobile Internet access (e.g., via Wireless Application Protocol WAP) is still limited in Germany, despite the extremely positive forecasts that accompanied its introduction. Studies such as that of the European Institute for the Media (EIM) on the World Internet Project (WIP) in Germany show that less than a fifth of all cell phone owners with Internet capability actually use mobile Internet services (Groebel & Gehrke, 2003).

Conversely, in Japan the mobile communications operator NTT DoCoMo and its i-mode service has established an extensive consumer base for its mobile Internet service. Forty million Japanese citizens (two thirds of the country's mobile communications customers) use i-mode. This is clearly reflected in NTT DoCoMo's revenues per customer. While subscribers to mobile communications services in Europe spend on average around € 400 per year (with a declining trend), in Japan i-mode users pay more than € 80 per month (IDATE, 2002, pp. 51–52, 62).

The Japanese example illustrates how the wireless digital future could look. With third generation mobile technologies, the consumer is offered a wide-ranging choice of multi-media communications services; voice telephony is just one among many. This new market model fundamentally alters the corporate value chain. Costs are no longer calculated on the basis of talk time but on the quantity and quality of data requested (cf. Friedrich-Ebert-Stiftung, 2002). Thus, the tasks facing the network provider are fundamentally different from when GSM was introduced. Alongside the infrastructure, it is important to build up a service and content portfolio, either working alone or in collaboration with other providers, as well as developing new business models and marketing strategies.

Assuming a virtually saturated mobile communications market, it is anticipated that business customers will prove most profitable initially, followed later by other groups in society. IDATE predicts a further increase in the number of cell phone users in Western Europe, the establishment of 3G, and an increase in per-head expenditure on mobile telecommunications services at the expense of pure voice telephony.

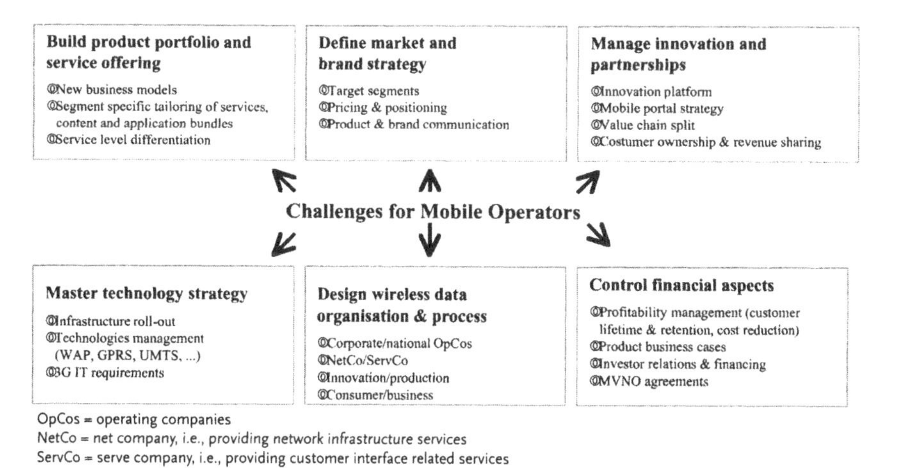

Figure 1: Wireless Data Challenges for Mobile Operators Europe
(Source: Booz Allen Hamilton & GCI Hering Schuppener 2001, p. 9)

Table 1: GPRS and UMTS market prospects in Western Europe

Evolution of the number of mobile data service users over GPRS/UMTS networks (millions)						
	2000	**2001**	**2002**	**2003**	**2004**	**2005**
Total number of cellular subscribers	243.0	275.8	299.2	316.4	328.6	337.2
Number of GPRS/UMTS users	0.0	0.1	3.7	18.5	44.9	81.7
Penetration	0.0%	0.1%	1.2%	5.9%	13.7%	24.2%
Evolution of the ARPU						
Voice ARPU (€)	39.8	35.0	33.0	31.8	31.1	30.9
Data ARPU (€)	2.5	2.5	4.0	4.9	6.4	9.0
Total ARPU (€)	42.3	38.4	37.1	36.7	37.5	39.9

ARPU: average revenues per user *(Source: IDATE, 2002, p. 66)*

But how realistic is customer interest in Europe? GSM fulfills the fundamental requirements of availability irrespective of location, simplicity and security. Deciding what will become the "killer application" of the up-and-coming mobile communications generation is a contentious issue. Mobile communications market analyst Mathias Plica, for example, sees "mobile bandwidth" as a true killer application (Pauler, 2002). Visualization by means of new services (MMS or innovative UMTS-services) may also prove a winner. The success of future mobile communications will be determined ultimately by the ways in which technical capability is used to address social needs.

Aside from the problems specific to each case, the underlying thinking is all too familiar. Ken Olson, founder and chairman of DEC, said in 1977: "There is no reason anyone would want a computer in their home." When making predictions about innovations, short-term developments are usually overestimated, and long-term effects tend to be underestimated. Technological developments and applications do not appear from nowhere; instead they are integrated as socio-technical systems into an evolving social and technological development process. Therefore, an analysis of key basic trends in technology and society provides an insight into general trends for the future.

3 Technological Development Processes: Basic Trends

The basic technological trends include miniaturization, digitalization, networking, and the further development of mobile communications structures.

3.1 Miniaturization

By miniaturizing electronic components one can increase the capacity and performance of IT and communications systems. Advances in microelectronics paved the way and provided the driving force for the dramatic overall growth in information and communications technology. One example is the Motorola International 3200. Known as "the bone" and weighing 520 g, it was one of the first GSM telephones back in 1992. A more recent Motorola device, the C 330 "hourglass", or Nokia's 8210 model weigh just 78 g and are much smaller, with far more functionality.

The technical integration of functions and portability of communications devices are a result of this process. Miniaturization and mobility go, quite literally, hand in hand. Cell phones featuring organizers, integrated digital cameras and navigation systems have been launched in the marketplace. Siemens has developed, for example, a "wrist phone" which is a miniaturized triple-band GSM cell phone. A device the size of a wristwatch contains voice-call functions, a hands-free device, triple-band GSM technology as well as an actual watch.

Through high data transfer rates and the further development of menus and displays, a smooth transfer from IP-based Internet services also becomes possible via UMTS-compatible cell phones (Gaida, 2001). The technological trend of miniaturization is experiencing a new spurt and is moving towards even smaller nanostructures with exciting new product innovations, not only in the area of microelectronics but also, for example, in the areas of biotechnology, medical technology, and materials technology.

3.2 Digitalization

It was the digitalization of all components in the communication process which first made it possible to dramatically increase capacity and converge networks, end products, services, and contents. The common denominator is zero or one.

Digitalization supports trends toward dematerialization and universalization: first, digital files can be copied and distributed as dematerialized products almost ad infinitum; and second, this makes digital files universally usable. The copy process replaces material production and physical transport is rendered unnecessary by digital data flows. The possibilities offered by currently available methods and technologies have not yet been explored (Fischer, 2002).

Moreover, dematerialization involves replacing products with services: for example, buying the rights to use a software program rather than acquiring and owning the product. Research using an online dictionary can theoretically replace the need for (and ownership of) a book. Physical presence is additionally supplemented via telecommunications services using virtual presence as an alternative. This further promotes the mobility and flexibility of the individual.

3.3 Networking

The third basic technological trend is networking. Use of networking for companies, individuals or groups increases through expansion once a critical mass of consumers is reached. Use of the telephone came into its own once it had reached a critical mass. Nowadays the Internet is the central catalyst for this type of network effect.

From a technical standpoint, networked cluster formation increases capability. In other words, the combination represents more than the sum of the individual capacities. Examples from the IT area are PC clusters or processor clusters. The same concept applies to the economy. Innovative company clusters can, for example, increase the economic performance of regions, with corresponding spin-off effects for the job market and for training opportunities, quality of life, etc. "A cluster influences the market in three ways. First, it creates greater efficiency (...). Second, it drives opportunities for innovation (...). Third, a cluster has a positive influence on the start-up rate of new firms", according to Harvard professor Michael E. Porter (Heuer, 2002, p. 21; Ketels, 2004; Steinbock, 2004). In Germany, an example for this type of company cluster is the IT cluster in Dortmund, North Rhine-Westphalia. The formation of networks and clusters will play a major role in the future of mobile telecommunications, the shaping of business models and the innovation potential of the sector (Steinbock, 2004).

3.4 Further Development of Mobile Communications Structures

Mobile communications increases spatial independence and offers greater room for maneuver. Higher transfer capacities of third generation (UMTS) or Wireless Local Area Networks (WLAN) allow a data-intensive and mobile visual communication (graphics, images, videos, music), which would have been unthinkable just a few years ago. Depending on the location and speed of the receiver as well as the number of users, transfer rates with UMTS reach between 144 kbit/s and 2 Mbit/s (Gaida, 2001, p. 62; Lehner, 2003).

Besides this increase in the performance and capacity of mobile communications structures, another major technological change in mobile communications is provided in the shape of "always on technology". 3G allows for flexibility and mobility in radically new ways, through personalized content at any time, in any place. A permanent online connection, the use of which is then charged not in terms of time, but according to the data packages transmitted, opens up fundamental new communication options (e.g., instant messaging, online games or up-to-the-minute transfer of information and news services).

4 The Social Context: Basic Trends

These fundamental technological trends are closely related to social developments, changes in users' interests and changes in the social 'climate'. This is all the more important given that specific individual technologies will—it is assumed—increasingly fade into the background. "Link-up technology will no longer be at the forefront as was the case for decades with analogue fixed network link-up technology (...). The infrastructure of the future will be more of a mix of different technologies, tailored to specific requirements." (OFCOM et al., 2001)

4.1 Globalization, Mobility and Flexibility

Globalization, mobility and flexibility are requirements of our modern society. "Always-on" strategies allow us to be contacted at any time and in any place. Permanent mobile communications is an essential part of life to many people and many suffer as a result: communication replaces physis.

Furthermore, new, flexible means of rapidly "switching" between private and business communication are available, dependent on context rather than location: Situation replaces localization.

This spatial flexibility is supported through "on demand strategies". Direct personal communication "outside the workplace" ensures a permanent connection to commercial or domestic information highways. Databases and information are available at all times and can be accessed whenever and wherever they are needed, irrespective of whether the person is in a particular location, either in the workplace or at home.

Whether such developments are actually a blessing or a curse is open to discussion. Anything can be found, but it is possible to get lost in the sheer volume of information available. The authors of the Japanese study for the aforementioned World Internet Project report, for example, found that Japanese citizens who use the Internet while on the move show a significant and marked lack of purpose regarding their objectives in life compared with those who access the Internet via PC or not at all. "Could this be a reflection of the group's absorption in communication with friends unsuccessfully finding their objectives in life?" (Institute of Socio-Information and Communication Studies & Communications Research Laboratory, 2001, pp. 132–133).

Do we have sufficient capacity to select and process information ourselves? Who is able to determine for themselves when to stop? Greater availability increases the pressure to perform and each individual has to make more and more decisions regarding what is really important.

At the same time, these new services allow a positive interpretation of globalization, flexibility and mobility. It is possible to call or transmit data to someone living in a wooden hut in the middle of nowhere. This promotes, as explained in the next point, teamwork and knowledge transfer.

4.2 Team Work and Knowledge Transfer

Comprehensive teamwork requires for the most part face-to-face meetings at selected locations, frequently involving travel. Electronic knowledge transfer provides an alternative through global network structures. Higher broadband capacities in the electronic transfer of information allow better visual communication and the integration of additional and contextual information (in-sight view) relevant to the decision-making process. In expanding and enhancing virtual communication content, we help satisfy one of the key requirements for fully functional knowledge transfer: the transformation of implicit personal knowledge into explicit common knowledge.

In private communications too, Multimedia Messaging (MMS) services allow freedom of expression and communication. Through MMS, images and sound can be sent independently of time and location. Mobile communications providers hope that, as a direct result of this type of visualization, the application will be widely accepted, thus generating greater opportunities for profit.

4.3 The Public-Private Relationship

Changes in communication infrastructures also allow conclusions regarding public/private interaction. Take, for example, the image of the telephone kiosk; telephone kiosks have been transformed from being fully enclosed into free-standing columns. "The world is a living room" (Rauterberg, 2002) ran the headline in the German weekly Die Zeit, "Cell phone boom. Telephone kiosks a thing of the past", according to Süddeutsche Zeitung (Bock, 2001). The same analogy, only more radical, can be applied to the use of cell phones in public places, whether this be a restaurant, train compartment or waiting room.

Irrespective of actual location, we switch happily between the private and public spheres. Communication can take place in any location and may be completely detached from the setting for the conversation (intimacy, grief, commercial secrecy). The public is given, often without any consideration, an "in-sight view" into private or commercial matters; this may be called push-privacy.

This has also proven to be the case on the Internet where it has become increasingly more acceptable to talk about private matters in the public arena. Intimate relationships, confidential matters and emotional private events are no longer kept discreet and instead are voluntarily brought out into the open via webcams, in chat rooms or online diaries; in this case a form of pull-privacy.

The increasing propensity to discuss private matters in public is an international phenomenon, encouraged partly by the population's desire to leave behind traditions and values passed down through the generations and to experiment with different media, but also through the efforts of media groups to produce content which is cheap and "up to the minute" (Koenen & Michalski, 2002).

The risks inherent in private exposure (e.g., infringement of privacy and human dignity as well as unintentional consequences of self-revelation), as well as new cultural impulses can be created which call into question long-standing norms and values through the public medium (e.g., social discourse on taboo issues such as illness, unconventional lifestyles etc.). Removing taboos in this way can give the impression that the virtual public arena is somehow less "threatening" than the immediate private environment. Unlike with mobile communications in public places, the net-using public is only confronted with such "private matters" if people actively search for them and access them "on demand" (Konert & Hermanns, 2002).

5 Visions for the Future and Outlook for "Always-On" and "On Demand"?

We are still some way from a world in which access to mobile platforms and applications is as simple as accessing electricity. The service is complicated and not exactly user-friendly. Accessing the Internet quickly via WAP or exchanging address and telephone details between two cell phones is something only die-hard users have the patience for.

There are many possible scenarios: Similarly to switching on an electrical appliance and getting electricity straight away, it should be possible to gain immediate access to the Internet without complicated menu procedures which vary from one service provider to another. Also, sharing digital data between technical end products such as digital organizers or cell phones could be simplified.

IBM, for example, has developed an idea which has now been adopted by the Japanese telecoms provider NTT and its mobile communications subsidiary DoCoMo. In the future, it will be possible to exchange telephone numbers or digital business cards electronically through a handshake (Ziegler, 2002, p. 50). Receivers are either PDAs or cell phones which automatically establish a small computer network through skin contact and transfer the required data using the electrical conductivity of the human body. The hardware need not be placed directly on the skin. It should be noted that the electrostatic fields generated are much lower than the charges you receive, say, when combing your hair—around 1000 times weaker. This form of interface between humans and machines could conceivably be developed as a biotechnical key for Internet access or other access rights, e.g. the door to one's house. Thinking even further ahead, by implanting technology chips in the human body, it might even be possible to remove completely the external technical interface in the shape of the cell phone or PDA.

When assessing the feasibility of these visions for the future, it is important to take into account not only technical capability but also the socio-cultural context. For instance, the body's own data transfer system described here might not be well received in parts of Asia given their cultural attitudes to human contact (handshake). The inertia effect of cultural context, ethical convictions and behavioral traditions are often underestimated by innovation-happy engineers.

The key to realizing future visions lies in combining technical capability and social desires. That is easy to say, but requires considerable effort on the part of all involved to lift their gaze beyond their own specialist world and carry out a realistic assessment of future developments and user interests.

References

Anonymous. (November 25, 2002). Studienmonitor: UMTS & Marketing 2002. Retrieved from the World Wide Web: http://www.wuv.de/daten/studien/072002/568/summary.html.

Ballon, P. et al. (2002). Business models for next-generation wireless services. *Trends in Communication (TIC)*, 9, 7-29.

BITKOM Bundesverband Informationswirtschaft, Telekommunikation und Neue Medien (Eds.). (2002). *Wege in die Informationsgesellschaft: Status Quo und Perspektiven Deutschlands im internationalen Vergleich.* Berlin: BITKOM. Retrieved from the World Wide Web: www.bitkom.org.

Booz Allen Hamilton, & GCI Hering Schuppener (2001). *Delivering on the Promise — Turning Wireless Data into a Success. White Paper — October 2001.* Retrieved from the World Wide Web: http://extfile.bah.com/livelink/livelink/103069/?func=doc.Fetch&nodeid=103069.

CIT Publications Limited. (2002). *Yearbook of European Telecommunications 2002.*

EMPA (n.d.). Dematerialisierung und Informationsgesellschaft. Retrieved from the World Wide Web: http://www.empa.ch/plugin/bean/empa/Article_PrintArticle?pr_artid=7041.

Fischer, H. (2002). Leute rauswerfen kann jeder. *Die Zeit, 26.* Retrieved from the World Wide Web: http://www.zeit.de/2002/26/Wirtschaft/200226_effizienztheorie.html.

Fiutak, M. (October 18, 2001). UMTS: Markt hat nach dem Hype Potenzial. *ZDNet News — Internet & Telekommunikation.* Retrieved from the World Wide Web: http://news.zdnet.de/story/0,,t101-s2097551,00.html.

Friedrich, J. (October 31, 2001). UMTS: Ist das teure UMTS überflüssig? [online] *Stuttgarter Zeitung.* Retrieved from the World Wide Web: http://www.stuttgarter-zeitung.de/stz/page/detail.php/48676.

Friedrich-Ebert-Stiftung (Eds.). (2002). *Die Zukunft von UMTS. Von der Begeisterung über Ernüchterung zur Belastung.* Gutachten: Arne Börnsen. Analysen der Friedrich-Ebert-Stiftung zur Informationsgesellschaft Nr. 7/2002. Electronic ed.: Bonn: FES Library.

Gaida, K. (2001). *Mobile Media: Digital TV @ Internet.* Bonn: mitp-Verlag.

Gordon, J. (2002). The Mobile Phone: An Artefact of Popular Culture and a Tool of the Public Sphere. *Convergence,* 3, 15-26.

Groebel, J. & Gehrke, G. (Eds.). (2003). *Internet 2002: Deutschland und die digitale Welt. Internetnutzung und Medieneinschätzung in Deutschland und Nordrhein-Westfalen im internationalen Vergleich* [Working Title]. Opladen: Leske + Budrich.

Heuer, S. (2002). Mehr Kunst als Wissenschaft. In: *McK Wissen,* 1, 20-25.
Hürlimann, A. (December 11, 2001). Und drittens kommt es anders, als man denkt. Mobilkommunikation zwischen Anspruch und Wirklichkeit. *Neue Züricher Zeitung.*
Retrieved from the World Wide Web: http://www.adlittle.ch/downloads/011211_NZZ_UndDrittensKommtEsAndersAlsManDenkt.pdf.
IDATE. (2002). Uncertainties in the Dynamics of Mobile Market. In: IDATE, Digi-World 2002. *The European way to think the Digital World* (pp. 49-69). Retrieved from the World Wide Web: http://www.idate.fr.
Institute of Socio-Information and Communication Studies, The University of Tokyo & Communications Research Laboratory. (2001). *Internet Usage Trends in Japan. Survey Report 2001.* Tokyo: Communications Research Laboratory.
Ketels, C. (2004). *European clusters. Structural change in Europe 3 – innovative cities and business regions.* Hagbarth Publications.
Knape, A. (2003). Wird UMTS ein Flop? Retrieved from the World Wide Web: http://www.manager-magazin.de/it/e100/0,2828,233021,00.html.
Koenen, A. & Michalski, R. (2002). Blick über die Grenzen: Transkulturelle Perspektiven auf eine globale Entwicklung. In R. Weiß & J. Groebel (Eds.), *Privatheit im öffentlichen Raum. Medienhandeln zwischen Individualisierung und Entgrenzung* (pp. 89-151). Opladen: Leske + Budrich.
Konert, B. & Hermanns, D. (2002). Der private Mensch in der Netzwelt. In R. Weiß & J. Groebel (Eds.). *Privatheit im öffentlichen Raum. Medienhandeln zwischen Individualisierung und Entgrenzung* (pp. 415-505). Opladen: Leske + Budrich.
Lehner, F. (2003). *Mobile und drahtlose Informationssysteme. Technologien, Anwendungen, Maerkte.* Berlin: Springer.
Mercer Managment Consulting, & HypoVereinsbank (n.d.). Medien-Studie 2006. Zukünftige Trends in der Medienlandschaft.
Retrieved from the World Wide Web: http://icfb.hypovereinsbank.de/pdf/ppt_6seiten.pdf.
OFCOM et al. (2001). *Konvergenz von Infrastruktur und Diensten der Fest- und Mobilkommunikation. Bericht über die Entwicklung der Technologien und Märkte sowie der sich dazu stellenden regulatorischen Fragen.* Retrieved from the World Wide Web: http://www.bakom.ch/imperia/md/content/deutsch/telecomdienste/telecominfomailing/Bericht_Konvergenz_d_07_02_01.pdf.
Pauler, W. (September 6, 2002). *UMTS — Start der dritten Mobilfunk-Generation. Die Handy-(R)Evolution.* Retrieved from the World Wide Web: http://www.xonio.com/features/feature_8755033.html?ly=print.
Rauterberg, H. (2002). Wohnzimmer ist überall. Über den Terror des Intimen: Warum immer mehr Privates zur öffentlichen Angelegenheit wird — und wie unsere Städte sich dadurch verändern. *Die Zeit, 3.* Retrieved from the World Wide Web: wysiwyg://11/http://www.zeit.de/2002/03/Kultur/print/_200203_tel.saeule.html.
Steinbock, D. (2004). *What next? Finnish ICT clusters and globalization.* Helsinki: Ministry of the Interior Finland.
Ziegler, P.-M. (October 21, 2002). Shake-Hands. Datenaustausch per Hautkontakt. *c't magazin für computer technik,* 22, 50.

3
On the Myth of Convergence

Klaus Goldhammer

1 Introduction

In many discussions on the future of media the term "convergence" is present, often uncritically acclaimed or prematurely held responsible for media developments. Yet the topic raises questions: What does convergence really mean? Where can we find true convergence? And if there is convergence—is it useful and valuable for the consumer? Such questions on convergence will be analysed and discussed in this article, focussing especially on the much heralded topic of converging wireless content applications.

Special attention will be given to three major areas: convergence of devices, content, companies, and markets.

2 Definitions

Convergence means that the realms of media, telecommunications and information technology seem to be conjoining. Definitions of the term come from different academic disciplines. Most remain imprecise—as for example the one outlined in the European Union's Green paper:

"The term convergence eludes precise definition, but it is most commonly expressed as:

- *the ability of different network platforms to carry essentially similar kinds of services, or*
- *the fusion of consumer devices such as the telephone, television and personal computer..."*

Probably the most precise description is provided by the mathematical definition: Two lines approaching each other infinitely close but never cross.

In this chapter, we will examine convergence in the wireless sector. At first, convergence of wireless devices will be surveyed.

3 Devices—Of Razors and Mobile Phones

3.1 The Issue of Practicability

The first issue is practicability of the converged device. By practicability we understand technological user-friendliness, i.e. the ease with which the different functions of the respective device can be utilized.

Consider for example the CASIO watches of the "e-data bank"-series offering a host of functions including a password-protected database and schedule planner. It certainly may seem useful to have access to important data at all times, yet trying to organize one's schedule on a watch display will surely prove a challenge.

Mobile phones offer some relief to that problem: their integrated schedule planners are considerably easier to use due to their larger displays. But their practicability is impaired by another problem: how to use the converged device to set a calendar event while having it pushed against the ear while making a call?

3.2 The Issue of Functional Efficiency

The second issue to be addressed deals with functional efficiency. Converged devices often do not live quite up to their promises. Consumer complaints[1] indicate, for example, that wrist-watches equipped with a host of tools such as a digital compass, an altimeter, a barometer, and a thermometer[2] often do not produce the same quality results for all the offered functions.

These two arguments do not negate convergence as such. Practicability could be increased and functional efficiency ensured. But as for the moment most of today's converged appliances do not meet consumers needs.

3.3 The Issue of Utilization Patterns

A third problem is the issue of device-specific utilization patterns. What all converged appliances have in common is that they combine formerly separate appliances into one, because the underlying tasks are assumed to be closely associated with each other, or accomplished at the same time.

Consider for example Internet-enabled TV. The reason behind combining the two functionalities of the TV and an Internet-enabled PC in only one device (Internet TV) was believed to be beneficial for the consumer.

Two radically different technologies were blended together, yet only little thought was given to the differences in their utilization patterns. Even though the technological foundation—the digitalization of content—could eventually enable the fusion of TV and the PC, existing utilization patterns are not likely to turn channel surfers into web-surfers. The fundamental problem of those devices can be described as the "Swiss Army knife dilemma" (Norman, 1998). While it surely is of great help to use a Swiss Army-knife outdoors, in daily life we prefer using a bread knife to cut bread, or a corkscrew to open wine bottles, to using our handy pocket knife indoors. Different utilization patterns demand different, specific devices. The reason is that only those tools, designed for a specific task can be perfectly fitted to that task in physical form, features, and structure. Whenever one device is made to fulfil several functions, it must compromise on how well it can handle each individual task. Furthermore, the increasing complexity is likely to scare off many people.

3.4 Why Convergence of Devices?

But if converged appliances create more problems then they solve, why do manufacturers keep launching them? Donald Norman (1998) argues, that the chief problem is engineers drive the high-tech industry rather than the needs of consumers. Norman accuses the industry of succumbing to the disease of "featuritis".

A second possible explanation are cultural differences. One example of this can be found by comparing the consumer cultures of the U.S.-American and Japanese markets, summed up in the term "mottainai" by Steve Mollman (2001). Roughly translated, it means "it's a shame when you waste something". This sense of "mottainai"—centuries old—leads to objects and devices in Japan being fitted together in intricate and complex ways, so as to save the most precious resource of all in Japan: space. Mottainai helps to explain the country's "converged appliance" phenomenon, and thus also accounts for the fact that it is often problematic to translate products successfully adopted on the Japanese market into the Western Hemisphere.

3.5 Summary Devices

"Our research has come to the conclusion, that even in 2010 we will not see a universal communication machine, that determines our everyday live" (Beck, Glotz & Vogelsang, 2000)[3].

Differences in the business models and price structures of wireless and conventional telephone networks should not be neglected. However, an economic analysis cannot explain which converged or specialized devices are being used for the respective network. Converged "do-it-all" devices do not offer the best solution for most consumers in Europe and North America.

4 Contents: Are We Content With Content?

In the field of content, the vision of a "converged future" is represented by the idea that content only has to be produced once, and can then immediately be sent on to a host of different platforms.

4.1 Conflicting Standards and Integration Issues Will Impair Progress

The medium determines the message. There are vast differences between what can be displayed on a TV-screen, a PC-monitor or a PDA-screen. First of all, a vast number of competing software and hardware standards confuses content providers and users alike. Much thought has been put into software that invisibly serves appropriate content to any device chosen. A step in this direction has been the extensible markup language (XML). Different standards are abundant for different applications (such as VoiceXML and ebXML (electronic business XML)). And even though several initiatives[4] try to fuse XML standards, the road towards universal convergence is still long if not impossible.

But as it has already been argued before—what might be technically possible is still a long way from what consumers will use and pay for. Regardless of available programming standards to make content accessible from a number of different devices—the tough question to answer is: Is it going to look good? And: Does anybody need it?

Display sizes, memory (RAM) and battery capacities as well as the speed of information transfer significantly limit content transfer from one device to another, even under the assumption of universal standards. Anybody who has tried to play an interactive PC-game on a mobile phone will agree (Screen Digest/Goldmedia, 2004).

4.2 Case Study UMTS

In the year 2000 several European countries auctioned licenses for third generation (UMTS) mobile telecommunications, the new standard that is expected to revolutionize mobile communications since it allows providers to offer a host of new mobile services, such as fast Internet access, virtual banking, credit card transactions, and video conferencing surpassing the quality of the most advanced fixed-line telephony.

The providers of mobile telephone services are now in a desperate need to find new meaningful applications that will help to recover the investments.

One of the first UMTS-applications that providers included in their portfolios was the videophone, representing the convergence of a camera and a telephone. We argue that it is more questionable whether the videophone will represent even a moderately successful application.

The idea for this type of device has come up many times already in history, reaching back until 1927 when Bell System manufactured the first "Picturephone", or a primitive trial version. At the 1964 World Fair in New York, Bell System unveiled a fully developed "Picturephone"-system. Not surprisingly, it never succeeded. *"The technology clearly was ahead of its time—[but] consumers and their lack of interest doomed the Bell System's Picturephone system,"* Communications professor A. Michael Noll (1992) wrote in an article titled "Anatomy of a Failure." The manufacturers were so convinced of its ultimate success that they even ignored their own market research findings, in which half of the people surveyed stated that they weren't interested in such a device, Noll explains.

Yet despite the past failures, the video phone concept has not deterred companies from trying again and again, as optimistic as ever that it will eventually succeed in a consumer market.

4.3 Driving Force for the Mobile Business

Far from assuming that product managers and companies are generally ignorant of what consumers need or even ignore their own research findings, it should be acknowledged that many companies conduct extensive surveys in order to be able to meet user demands and expectations. German mobile phone manufacturer Siemens (2001) for example asked 11,000 European users of mobile services for their requested services and potential willingness to pay for extra service.

Siemens found that 88% of the interviewees requested expanded mobile services. Especially the younger segment of the population turned out to be

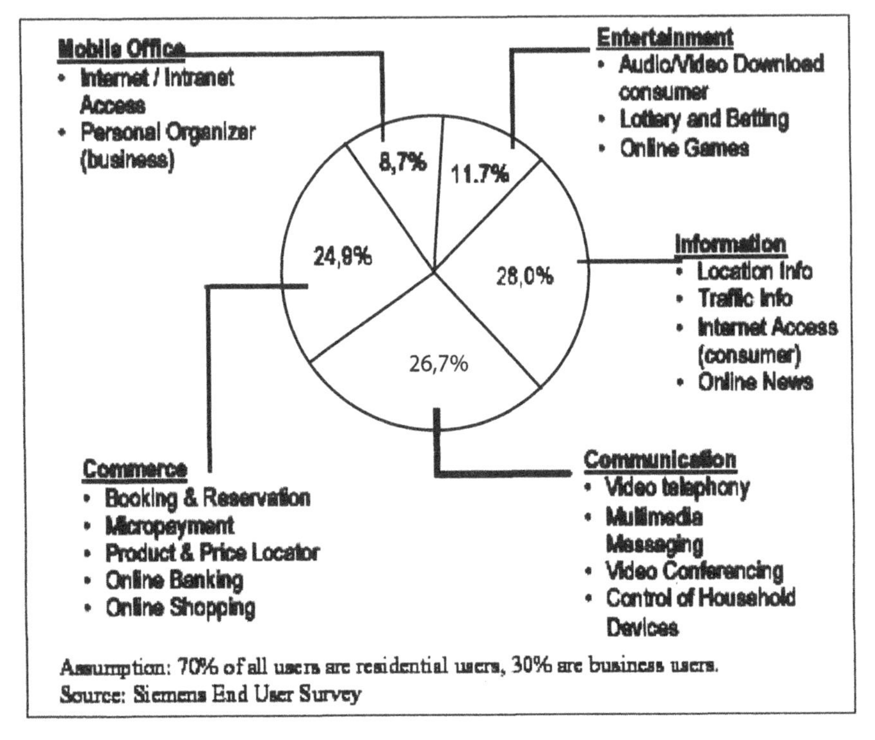

Figure 1: Requested services and user demand of mobile services

willing to pay for additional services. This so-called "generation@", represented by the under-25 user segment of the population, shows more interest in entertainment applications such as mobile interactive games or downloading audio/video files. Therefore, it is at the center of Siemens' marketing initiatives.

These findings are supported by a study conducted by the German B.A.T. institute (2002)[5]. In this survey, 2000 people were asked whether they would prefer an all-in-one device that combines telephone, TV, PC, Internet and e-mail functions. Not surprisingly, the population segments of the 14 to 19 year-olds as well as the 20- to 29-year-olds showed the highest interest compared to other age groups. As illustrated in figure 2, the demand for a converged "all-in-one" device decreases with advancing age. One reason might be that such young people have not yet developed distinctive habits of media usage for print media, radio and TV.

But how can the so-called generation @ be turned into the driving force for mobile business? Unfortunately, even though young population segments show a willingness to pay for additional services, they also have rather low levels of disposable income.

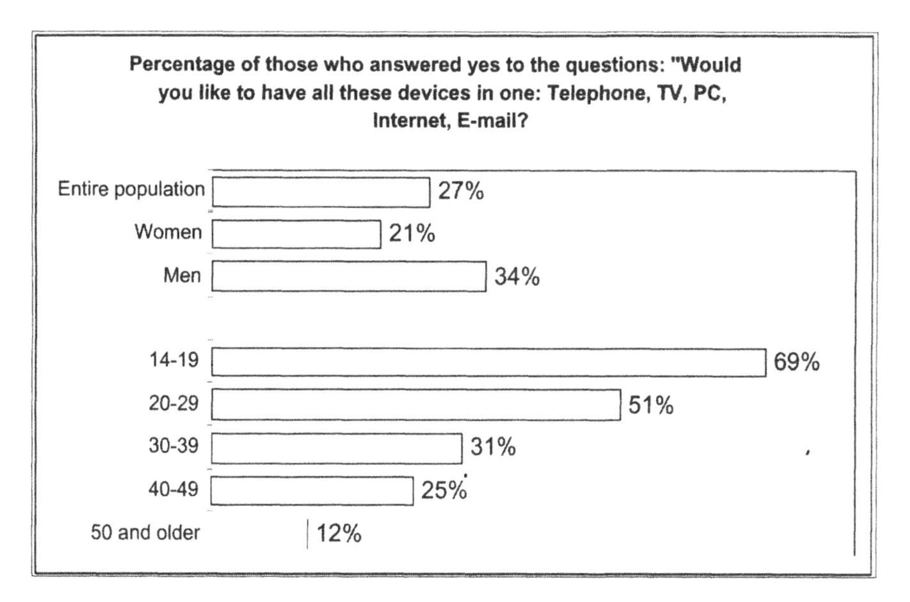

Figure 2: New media technologies vs. old media habits
Representative survey of 2.000 adults (age 14 and older), February 2002 in Germany (Source: BAT, 2002)

Which elements of mobile services will turn out to be successful? In its survey, Siemens asked respondents to indicate their willingness to use and to pay for specific mobile services. Downloading video/audio files is of high interest to 67% of the under-25 user segment. On average, these users are willing to pay € 3,20 per month for this service. Considering all age groups, though, only 38% are interested. The overall user group is only willing to pay an average of € 1,40 a month.

Receiving information seems to be of high interest to the average user. Survey data suggest that a map-based traffic information service for cars, would generate much interest. Users show high willingness to use it (68%) and are willing to pay about € 2,90 a month for it.

4.4 Content Delivery Services

Since 1992, when the standardization of GSM (Global System for Mobile Communication) was completed, GSM has become the most common technology for mobile telephony. In addition to speech, GSM offers a wide range of supplementary services such as call forwarding, call barring, multi-party service, and Short Message Service (SMS).

SMS was introduced as a service to notify mobile service subscribers of new messages in their voice-mail box. Later, person-to-person messaging became available, but only when cross-network text messaging was introduced the popularity of SMS increase drastically. As more mobile customers adopted SMS, a number of SMS-based mobile content delivery services were introduced. Users could receive information such as weather or sports news on their mobile phone.

The bulk of all messaging traffic is made up by person-to-person messaging. Specialized SMS content delivery services continue to be developed and will account for 7 per cent of all SMS traffic by 2006 (Frost & Sullivan, 2002). Network operators and content developers use the format as a support for next generation multimedia applications and services like EMS[6] and MMS[7].

4.5 Divergence not Convergence of Content

At the moment, content providers have to face the problem that there is no such thing as a write-once-run-anywhere content in the digital world. But as the survey of recent technological developments shows—for example, the convergence of standards, the increase of transmission bandwidths—convergence of content services to be used on a number of different devices, is probably going to be possible soon—technically.

Yet, user-friendliness is not achieved by technological convergence alone. High-resolution graphics do not translate well to a hand-held computer. Even a short text text on a PC-screen will lead to extensive scrolling on a mobile phone display. Thus, whether converged content will be visually suitable for all the devices it can technically run on, still remains an open question.

No matter what medium is used, it will always be necessary to format the content specifically in order to fit the method of interaction of each device and its technical specialities. The entertainment-format "Who Wants to Be a Millionaire" looks very different on a PDA and a PC.

A popular application on the PC and mobile phones alike is playing games. But trying to translate PC-games to run on a mobile phone can result in a less than enjoyable experience. Only one of the pictures can be displayed at each moment because of the limited display-size. Furthermore, the information transfer is slow, the displays are too small, memory (RAM) is undersized and the batteries are short-lived by coloured displays of mobile phones.

The market for wireless games has an enormous growth potential, and should reach $6 billion in 2005 in the U.S. and Western Europe (UK and Germany as the largest wireless gaming markets). But initially, only simple games such as card and quiz games and bingo were really suited and popular on mobile phones.

4.6 Summary Content

Visionaries are still far ahead of technological reality. And even though technology might soon make some visions come true, it is the users who decide in the end. For them, cutting-edge technologies count for less than content formats designed to meet their needs.

5 Companies and Markets

For quite some time, the convergence of three different markets has been heralded: Media, IT and Telecommunications companies were about to converge, creating a vague but promising "interactive multimedia world". And indeed, some companies tried to make this forecasts come true and to expand their markets. The most prominent examples are probably the merger of AOL and Time Warner, or of Spain's Telefónica buying Endemol. Many others were pressing ahead with take-overs, joint ventures and alliances along the value chain.

But after the dust has settled, what is the real result of the promised convergence of three markets?

The AOL/Time Warner merger attracted many investors because of their belief in convergence and its promise of the synergies. One of the chief intentions was the introduction of broadband Internet via Time Warner's cable networks. As a result, AOL would get access to content from TV channels and movie studios. In return, Time Warner could expand its Internet presence with the help of AOL.

These synergies did not materialize. On the contrary, the only definite result of this merger is a record destruction of values. While at the time of the merger in January 2000 the companies had a combined value of $181 billion, AOL Time Warner was worth only $106 billion in January 2001, after the merger was finalized. In 2002, the company was forced to take an accounting write down of $54 billion, a record amount.

6 Summary and Conclusion

After surveying forecasts and realities we argue that the empirical evidence does not support many of the scenarios of convergence in media, telecommunications and information technology.

The great majority of users do not want a universal device which combines a large number of different functions. Most consumers are not ready to change their habits and accept new (converged) systems without obtaining a clear added value.

We argue that product differentiation, divergence, and single usage applications will succeed. The continuing need to build device-specific content as well as missing standards show that the term "convergence" is vague and possibly mythical.

Whatever the focus is, be it devices, content or companies and markets—true convergence is rarely seen. Many visions lost focus of the consumer and his needs. Instead, the search is open for more realistic and concrete solutions.

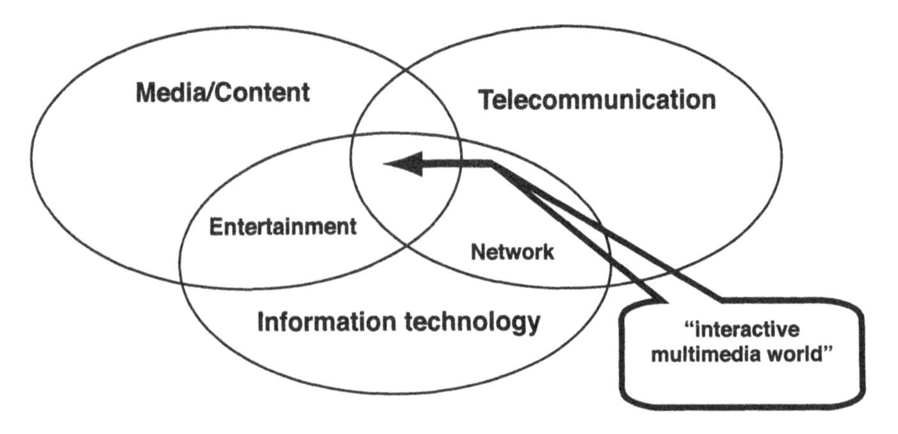

Figure 3: Market convergence

Endnotes

1 http://www.ciao.com.

2 These functions are offered by the CASIO Protrek Triple Sensor PRT-410T.

3 "*Unsere Untersuchungen legen den Schluss nahe, dass auch 2010 nicht die universelle Kommunikationsmaschine den Alltag prägen wird.*"

4 One of these initiatives is RosettaNet—an industry consortium helping to steer XML's development. It has more than 400 members including Cisco Systems, Microsoft, Intel, and Hewlett-Packard. Further information can be obtained from: *http://www.rosettanet.org.*

5 B.A.T. Freizeitforschung is a social research institute focussing on the use of leisure time. It was founded and is financed by British American Tobacco.

6 Enhanced Messaging Service: An advancement of SMS, that will lift the 160 symbol-border by linking several SMS.

7 Multimedia Messaging Standard: An advancement of picture and text transfer in mobile networks via SMS. MMS will be enabled by increased mobile-bandwidth.

References

Boston Consulting Group. (2002). *Mobile Commerce: Winning the On-Air Consumer.*

British American Tobacco/Freizeitforschungsinstitut. (2002). B·A·T Medienanalyse 2002: Wer will die neuen Alleskönner? *Freizeit aktuell, 166/23.*

Frost & Sullivan. (2002). *Frost & Sullivan's Analysis Of SMS Applications and Services in Western Europe (Report B056).*

Mollman, S. (2001). Japanese design sense becomes more evident in high tech products. *J@pan Inc Magazine—The J@pan Inc Newsletter,* 134. Retrieved from the World Wide Web: http://www.japaninc.net/newsletters/?list=jin&issue=134h.

Noll, M. A. (1985). Videotext: Anatomy of a Failure. *Information & Management, 9/2,* 99-109.

Norman, D. A. (1998). *The Invisible Computer.* Cambridge: The MIT Press.

Office of Telecommunications (2002). *Consumers' use of mobile telephony Q8.* Retrieved from the World Wide Web: http://www.oftel.gov.uk/publications/research/2002/q8mobr0402.pdf.

ScreenDigest & Goldmedia. (2004). *Wireless Gaming: operator strategies, global market outlook and opportunities for the games industry.* London/Berlin.

Siemens (2001). *Information and Communication Mobile: Whitepaper UMTS.* Retrieved from the World Wide Web: http://www.siemens-mobile.com/pages/monaco/include/pdf/whit_d.pdf.

4
Automotive Telematics: Is it Time for a Renaissance or an Obituary?

Jonathan Lawrence

1 Introduction

Recent years have seen a rapid boom and bust cycle surrounding the automotive telematics market in the United States. This cycle is the result of two primary forces: the prevailing overall economic conditions in the U.S. technology sector and the inability of the market to create a successful business model to serve as an example and foundation for further growth and investment. While telematics services and applications have been available since the mid-1990s, the mainstream emergence of automotive telematics was not generally recognized until GM OnStar's mass launch in 2000, when the boom cycle in venture capital for telematics-focused companies was already at its peak (Barraba et al., 2002).

Although there has never been a generally accepted definition of automotive telematics, the buzz surrounding perceived market opportunities grew to a fever pitch just as the U.S. technology bubble was beginning to burst. Just between January 1999 and April 2001, it is estimated that private companies who focus primarily on telematics market opportunities received nearly $1 billion in venture capital (Figure 1). This excludes investments made directly by automakers in subsidiaries (e.g., General Motors' OnStar, Ford's Wingcast joint-venture with Qualcomm) and investments made in public companies.

Automakers were scrambling to put together telematics ventures believing that the automobile was the final frontier in wireless communications, a destination for mobile content delivery that they could control, and a potential cash cow. After all, most individuals spend the overwhelming majority of their time in their home, their office or their car. Automakers focused on the fact that 50%-70% of all wireless airtime is used in the vehicle. Finally, they saw the business of automotive telematics (herein defined as vehicle-related communications and services) as a potentially large, non-cyclical, high-

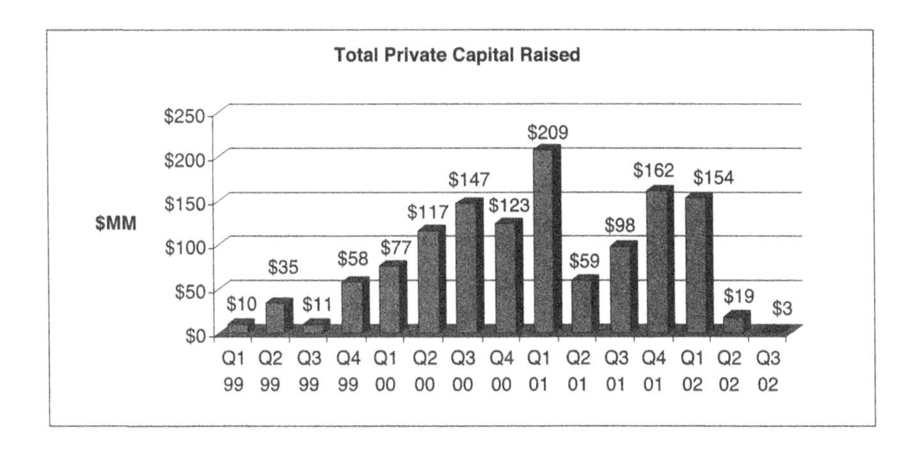

Figure 1: A Brief History of Venture Capital In Automotive Telematics
(Source: VCBuzz and corporate reports)

margin revenue stream for the otherwise intensely cyclical, low margin auto industry.

The wireless industry agreed and joined the automotive industry in creating business plans and making investments to address the opportunities that some analysts estimated could reach $50 billion a year worldwide by 2010. However, both industries did so largely independently of one another and soon realized their views and goals were not compatible. And so the market developed based on some false premises.

The aim of this chapter is to examine the development of the automotive telematics market, its current dysfunctional state, and to suggest a new paradigm for the future. It begins by briefly reviewing the misconceptions that led to inflated market forecasts, misguided business models, and over-investment. Then analyzes the dynamics between various links in the value chain in an effort to illustrate the types of inter-industry collaboration necessary to bring telematics applications and services to fruition so that market opportunities can be successfully realized.

2 The Origins of the Automotive Telematics Market

General Motors and Ford Motor Company began offering enhanced in-vehicle communications options in the mid-1990s. At that point, 'enhanced' simply meant any solution more sophisticated than an embedded cellular phone. While Ford's solution and plans for telematics remained modest, its rival GM, perhaps inspired by the torrid growth of its Hughes Electronics subsidiary which operated DirecTV, the nation's largest satellite television service, began planning for an all-out assault on the market for mass media content for mobile wireless communications in the automobile.

GM's OnStar subsidiary has since grown into the nation's largest telematics service provider (TSP) with three million subscribers in 2004 (OnStar, 2004). Still, OnStar has yet to turn a profit and legislative developments in the wireless industry have all but eliminated the possibility of that occurring for years. While OnStar continues to operate and even win new automotive original equipment manufacturer (OEM), its business model is generally regarded as unsuccessful. Based to a large extent on OnStar's results, Ford decided in 2002 to scrap its own telematics joint venture Wingcast, whose blueprint followed the OnStar business model. Ford had already plowed several hundred million dollars into Wingcast, along with the wireless technology company Qualcomm.

3 False Assumptions, Hard Truths, and Lessons Learned

Next is a review on the primary issues and challenges that have impacted the development of the automotive telematics market.

3.1 Embedded Wireless Technology

In retrospect, the most glaring mistake automakers made, and continue to make, is the decision to embed non-scalable/non-upgradeable wireless hardware in vehicles. 95% plus of the telematics enabled vehicles on the road today use an embedded analog only transceiver as the communications link to enable service. The theoretical problem with embedding wireless technology in automobiles is that technology changes rapidly, while automobiles stay on the road for 15-20 years and can only be modified or retro-fitted at significant cost. The inability to predict or adapt to future changes in technology makes embedding wireless communications technology in vehicles illogical.

As an example, in 2002, the FCC agreed to phase out the analog wireless network compatibility mandate (i.e., that all cellular handsets sold in the U.S. also work on existing, but outdated, analog wireless networks) on all carriers over a 5-year period. The wireless industry pressured the FCC to drop its requirement in order to faster upgrade to digital networks and improve spectral efficiency. Both OnStar and ATX Technologies (the two largest TSPs) vehemently opposed this action. The net effect of this ruling on TSPs and the OEMs whose vehicles they serve was to effectively place a 5-year 'expiration date' on all existing subscribers. At yearend 2002 there were approximately 8 million vehicles on the road with embedded analog-only telematics technology at an original equipment cost of over $3 billion. By 2008, the value of this investment (excluding R&D and cost of service) should fall to virtually nothing. The cost for an OEM to retrofit an existing subscriber's vehicle are estimated to be at least $700, making it an unattractive option. High churn rates and vehicle turnover should erode most of the existing subscriber base by 2008 anyway. Nevertheless, the nature of the auto industry and the difficulty of changing production lines and vehicle designs means that the majority of automotive telematics systems being factory installed after 2002 for quite a while remained analog-only, despite the fact that automakers knew that analog networks will largely be gone by 2008.

3.2 The Perils of Wireless Resale Business Models

Compounding the obsolescence of the technology are the inherent flaws in the business models. TSPs buy large blocks of wholesale airtime to provide basic services and subscribers usually have the option of purchasing more pre-paid airtime for premium services. Thus, TSPs operate wireless resale business models. These models historically have failed because margins on the mark-up of airtime that resellers are able to garner tend to be razor-thin—a dynamic that the wholesaler controls indirectly. Because the carrier dictates wholesale airtime prices, they can ensure that the price of the reseller's service (no matter how differentiated it is) never becomes compelling enough to lure their more profitable direct subscribers away. This is true of TSPs, where carriers already offer the majority of premium services (non-vehicle specific) themselves. Consumers will not pay twice for the same services their wireless carrier already provides.

So TSPs and automakers are now focusing their business models and marketing on the safety and security services (e.g., automatic collision notification or ACN) that they alone can control. This brings us to another false assumption that is typically made in early stage markets, and was certainly made in the case of automotive telematics—accurate market forecasting.

3.3 Accurate Market Forecasting

The most common misconception relating to automotive telematics is that it is an industry. It is not. Automotive telematics is a set of applications and services based largely on existing technology that deals specifically with the automobile. Many of these applications and services represent new revenue and cost saving opportunities for the wireless, mass media content and automotive industries. The natural evolution of the wireless industry has created many offshoots similar to telematics, including location-based services (LBS) and mobile commerce. These too have been viewed as distinct 'industries'. A cursory examination of the major themes of these market opportunities reveals that they largely overlap in the products, services and applications they encompass. Thus, there is a great deal of double and triple counting of opportunities that has left the market with wildly optimistic forecasts for growth and demand.

3.4 Safety Sells—But at What Price?

Automakers conducted numerous studies and focus groups to determine whether and how much consumers would be willing to pay for various services, including automatic crash notification, access to news reports, stock quotes, and sports scores and real-time traffic reports to name a few. In some studies, consumers were asked to rate the importance of more than a dozen service offerings and how much they would be willing to pay for each. Of course, different services are of higher value to different people. However, *these services cannot be purchased individually* and the failure to recognize the implications of this on consumers led to a myriad of market forecasting errors. The business practice shows that TSPs bundle services together into "packages". Consumers are smart enough not to pay twice for the same services and when confronted with the option of purchasing a bundle of services, some of which they already have, don't want or don't need, they have opted not to purchase at all. Unfortunately, automakers sell vehicles mostly from inventory through dealer networks. So they install all of the hardware necessary to enable all services, before knowing whether the eventual buyer will ultimately pay for any.

While consumers seem most interested in safety and security services, they don't value them highly enough to justify the costs associated with installing them in a subscriber-oriented business model. Low retention rates have proven this to be true. Most consumers ultimately expect to receive these services for free, and that is likely to happen beginning with luxury vehicles, as manufacturers attempt to utilize telematics as a product differentiator in an increasingly competitive selling environment.

3.5 Subscription based Telematics Business Models

Mercedes, BMW and other luxury manufacturers, some of whom offer OnStar, price-impact their vehicles for telematics systems or offer them as dealer-installed options. Either way, they are recovering the costs of the system (and usually the first 2-3 years of service) at the point of sale, greatly reducing, if not eliminating any dependence on the consumer to become a permanent subscriber. Luxury manufacturers can do this as their customer base is far less price sensitive. For example, BMW only offers their 'Assist' telematics system standard on their most expensive models, the nearly $80,000 7-series. The system is available as a dealer-installed option on lower-priced models (~$4,000 for the full system before installation costs), but the option take rate is under 10% according to a random sampling of BMW dealers who were surveyed. BMW makes money by *not* offering the system standard, only to those that can afford it and are willing to pay for it up-front on a cost plus basis.

GM OnStar, on the other hand, operates a subscriber business model. Most other volume manufacturers largely decided to wait to see OnStar's results before making their own automotive telematics plans. Following OnStar's mass roll out, Ford created a joint venture with Qualcomm (Wingcast) and proceeded to largely re-create the OnStar blueprint. Roughly a year into the project, which was behind schedule and over budget, Qualcomm pulled the plug on their end of the venture. Ford was left to fund Wingcast's launch itself, a cost they were unwilling to bear after reviewing the first two years of results in the telematics marketplace. Ford dissolved Wingcast in 2002. OnStar is the only purely OEM-operated, subscription-based, volume TSP in the U.S. Nearly all of ATX Technologies' subscribers are under multi-year pre-paid contracts with the manufacturer (e.g., Mercedes Tele Aid) most of which have not yet come up for renewal.

In the U.S., OnStar and ATX Technologies dominate the TSP market. OnStar uses a subscription-based model, while ATX operates as an outsourced service provider. ATX managed to turn cash-flow positive around March, 2002 with fewer than 400,000 subscribers according to company officials. GM officials admit that OnStar did not turn a profit despite millions of subscribers. However, the comparison is not an apples-to-apples one. OnStar's business model includes hardware and marketing costs, while ATX is an outsourced, private-label service provider that operates without these costs. The important difference is the OEMs' approach: If the OEM recoups hardware and service costs up-front, telematics can be a profitable business. If not, profitability becomes incredibly difficult and the only justification for offering service is to have a competitive product in the marketplace to help maintain, or even grow market share.

4 The Business Model of the Future

4.1 Focus on Core Competencies

Analysts, investors and market participants are looking at the current state of disarray in automotive telematics and reaching the conclusion that the telematics business model enigma must be far more complex than previously thought. In contrast, I believe that market participants have been overcomplicating the business model. The existing business models for the industries that are needed to enable automotive telematics are already well understood. The wireless telecommunications and automotive industries need to adapt so that they can work seamlessly together to bring automotive telematics services to consumers. If each industry can resolve itself to focusing on its core competencies, there are benefits to be reaped from the opening of new revenue streams, the enhancement of existing revenue streams, accelerated cost reductions, and improved customer satisfaction.

4.2 The Automotive Telematics Value Chain

The telematics value chain consists of three broadly categorized industry groups: the wireless industry, the auto industry and the mass media content and service industry. Mass media content and services is the most loosely defined of these groups, as we include automated information services, voice-enabled Internet content, live operator assisted services, and emergency services.

Existing mass media services have proven to be the most difficult to bundle and sell to consumers for many reasons. Most importantly, as mentioned above, the majority of drivers simply don't need, don't want, and are unwilling to pre-pay for these services in addition to their existing cell-phone bill. There is only a small minority of drivers that includes business professionals, business fleet owners, affluent consumers and "gadget enthusiasts" that make up the legitimate target market for delivery of mass media content to the automobile. Second, consumers can already access much of this content, in particular wireless information services such as stock quotes and sports scores, through their existing cellular phone service. Wireless carriers continue to expand the breadth of information and services offered on wireless devices as handset technology and wireless networks improve in order to keep existing subscribers and attract new subscribers. Third, in-vehicle telematics user interfaces are prohibitively expensive, inevitably obsolete, not scalable and cumbersome to manipulate. From OnStar's voice-activated, embedded hardware system starting at approximately $300 (basic service starting at $200 per year) to basic navigation displays costing $1,500 to fully-integrated, top-of-the-line systems such as BMW's complete Assist package at over $4,000, automotive telematics user-interfaces are several times more expensive than the average cell-phone, while their installation in the vehicle prohibits scalability.

We have left automotive OEM-owned, 'end-to-end' TSPs out of the value chain as they can be disintermediated by other segments of the value chain in most of the functions they serve. OEM-TSPs have argued that the centralization of various services and their ability to manage and deliver content specifically for the vehicle environment makes them indispensable. This is only true if automakers refuse access to vehicle systems and the related information and services that can be offered (e.g., ACN). OEMs guard this information closely, especially diagnostic codes. Until 2002, automotive OEMs refused to even agree to divulge to independent repair shops the diagnostic codes and tools necessary to perform many repairs on their vehicles. The issue became so politicized that legislation was introduced in the Senate to force automakers to change this practice. In September 2002, automakers finally agreed to comply with new rules granting repair shops access to the codes by August 31, 2003. This battle demonstrates how valuable this 'content' is to automakers. However, automakers have no experience in delivering this content as a service provider.

4.3 The Auto-Maker as Content Provider Instead of Service Provider

Auto OEMs should abandon the service provision business in favor of being content providers and service enablers. Since auto OEMs own the "real estate" in the vehicle, they can establish a defensible market position in selling access to vehicle-specific content and service provision. They can do this without embedding expensive, non-scalable hardware. The EPA's OBD-II (on-board diagnostics phase II) emissions monitoring guidelines, posted on the EPA's web-site, mandate the centralization of vehicle emissions data and many OEMs have expanded the functionality of these systems to include diagnostic information and control over a multitude of additional vehicle systems and sensors. Just as auto OEMs recently agreed to give independent repair shops the information needed to access these systems, they could agree to provide access to other companies for a fee.

As a content provider, OEMs could get immediate access to the huge number of wireless subscribers in the United States today. Of course, not all vehicles on the road today are OBD-II compliant, as the mandate went into effect for new vehicles in 1996. Nevertheless, the greatest benefits in telematics for OEMs lie in obtaining diagnostics data on new model introductions to reduce the number of recalls and their associated expenses. Additionally, OEMs would not need to set up or manage a billing system and other back-office operations. This is not an area of core competency for automakers. Consumers are more likely to pay for incremental automotive telematics services if they are bundled in with their existing wireless bill. This model is also more conducive for pay-per-use billing in that consumers would be able to pay for just those services that they wanted and on the same bill they receive from their existing carrier.

4.4 The Benefit of Long Vehicle Lifecycles

One of the most frequently cited challenges in automotive telematics is hardware scalability. Automakers and wireless carriers understand the challenge but there are benefits as well. While one cannot predict the course of technology over time, we do know that the vehicle's systems will stay the same over its lifespan. Thus, wireless device manufacturers need only to make sure that their next generation products are backwards compatible with the vehicle's technology in the same way that Microsoft introduces new versions of their software that are compatible with older versions. This does not always work perfectly, but the point is that the vehicle's interface with the wireless communications device it utilizes should be kept as simple as possible. The vehicle bus can serve as the central "database" for vehicle systems information. The data should be accessible through existing standards, ones that would be relatively inexpensive to maintain if they became obsolete. This would not preclude encryption of the data which automakers consider critical to guarding their product secrets.

As a content provider, auto OEMs could work with any carrier the customer chooses. Since carriers subsidize a healthy percentage of the cost of wireless handsets, carriers would also be willing to subsidize part of the incremental hardware costs to enable automotive content access. This is a simplistic overview of just one potential solution for unlocking the value of telematics, and it is not without significant challenges. Nevertheless, it is an example of the kind of role each player in the value chain needs to resolve themselves to play if the value of telematics is to be unlocked without help from the public sector. The public sector's role here is undefined and complicated and largely gets into work in Intelligent Transportation Systems (ITS) which are at this point unrelated to telematics and certainly unrelated to mass media.

5 Summary

The market for automotive telematics applications and services has developed in a dysfunctional state, based on some false assumptions and the inability of diverse industries to work together successfully. The overwhelming majority of existing telematics enabled vehicles should have a limited useful life, due to the FCC's ruling to lift the mandate on wireless carriers to maintain analog compatibility in their networks. The cost to upgrade the necessary in-vehicle hardware is prohibitively high. Subsequently, it is not feasible to place any long-term value on the current pool of automotive telematics subscribers in the U.S.

While market participants have learned many lessons, new business models have been slow to emerge. This is due to the economic environment, long automotive product cycles and the inability of technology companies and automakers to successfully collaborate with one another. Companies in the automotive, wireless and mass media industries need to recognize their place in the value chain and resign themselves to delivering products and services within the scope of their core competencies. This means that auto-makers need to transition their role as service provider to one of service enabler and content provider, leaving service provision and the majority of mass media content delivery to wireless carriers and media conglomerates.

References

Barraba, V., Huber, C., Cooke, F., Pudar, N., Smith, J., & Paich, M. (2002). A multimethod approach for creating new business models: the General Motor's OnStar project. *Interfaces, 32*(1), 20-34.

OnStar. (2004). Onstar reaches three million subscriber milestone. Retrieved from the World Wide Web: http://onstar.internetpressroom.com/pressroom.cfm.

II

Content Models

5 Are There Content Models for the Wireless World?

Benjamin M. Compaine

1 Introduction

Wireless is an exciting, hot topic. Wireless is old news.

Both statements are accurate. Wireless is indeed an exciting topic, for users as well as for providers. First, cordless telephones liberated us from sitting by a desk or standing in the kitchen to speak on the telephone. Then, cellular telephones freed us to speak, send, and receive calls from almost anywhere. Hundreds of millions of subscribers signed up in the last decade. Newer technologies, under the labels 3G and WiFi, are delivering the same untethering of data as the previous technologies provided for voice.

My focus here is not technology, but *content models*. This is another name for a business model that reflects the revenue created by content in the value chain. This is different than models dealing with the equipment or the sales of wireless network services themselves (e.g., monthly cellular subscription fees). The question is therefore: "How can content providers make money delivering their wares via a wireless process that goes beyond their known business models?"

The old news is that radio and television have been delivered via a wireless format for decades. The business models for these processes are well known: some advertiser, some government supported, some in the form of direct user funding. So we indeed know the parameters of several content models for wireless.

Nonetheless, that would be the easy way out of this. This chapter will go beyond the traditional wireless models of broadcast to consider more complex ones: "How will the wireless telecom providers and the providers of content take advantage of the mobility afforded by the new wireless networks to generate greater revenue?" "How will the revenue generated be split between service and content providers?" Or "Assuming the technologies and economics come together sometime soon—and they will—what opportunities does wireless hold for those with content to sell?"

In short, "How can mobility add value to content?

If history can be helpful, the lessons drawn from the first wireless services should be instructive. It was not obvious to the early radio broadcasters what the successful models for the radio business would be. Some broadcasters in the 1920s in the U.S. tried asking listeners to subscribe or send in payments on a voluntary basis.[1] Westinghouse, which had a stake in the equipment business, supported its own station, as did AT&T. Only gradually did the advertiser supported model gain adherents. And that model was so-called sustaining programming. That is, the sponsor often owned and created the program. In the U.S. this model carried over into the early days of television. Gradually, the model evolved into networks and unaffiliated production companies creating the programming, with advertisers generally buying individual commercial slots rather than sponsorship.

The major alternative model for the form of wireless we call broadcasting was the public service model. In this case, taxpayers or listeners, directly or indirectly, funded the content. In Great Britain, a license fee on radio and television sets effectively created a user-supported system. In the U.S., a public service TV network was partly taxpayer supported, with the balance provided by voluntary viewer contributions and corporate "underwriting" of specific programs.

All this is to illustrate two points about wireless models (and print as well): First, business models do not necessarily spring up mature and full-blown. They often evolve. Various models may hold promise but give way to others that better fit a culture, an economy, a regulatory structure, or the usage patterns of consumers. Second, there may be no single model that always works. Rather, several models may be adopted simultaneously.

There are, then, models we know about that will work for some providers, some of the time. There are other models subject to speculation which may or may not work: we would only learn if someone tried them. Indeed, the so-called "dot com" bubble was a wonderful era of experimentation. Private investors risked their capital with the hope—in retrospect perhaps blind hope—that they would be part of a successful business model. Some were, in fact, successful: eBay found a business as the intermediary for auctions, taking a small percentage of each sale. PayPal grew in response to the success of eBay, providing a payment mechanism for individual and small merchants. *The Wall Street Journal* found that their brand name and the type of content they provided could attract paying customers as well as advertisers interested in reaching them.[2]

Many others, however, learned that neither advertisers nor consumers were willing to pay enough to cover their costs of gathering, creating and publishing their content online. Still others found online a useful adjunct to their business, either for promotion or direct sales. Some sellers, Amazon the most visible, created a direct sales organization entirely based on access of consumers on the Internet.

Out of this still ongoing experimentation with business models we have learned that in the Internet world the are few models that are proven templates. A model that works for one provider, e.g., *The Wall Street Journal,* does not necessary work well for another, e.g., *Slate.* Thus, it should not come as a surprise that models for what may work for content in a mobile wireless world will not be definitive and predictable.

For example, the i-Mode platform of Japan's DoCoMo wireless system is widely considered to be the only successful mass audience wireless system with a substantial content provider component. In 2002, there were more than 33 million i-Mode subscribers, two-fifths of whom had Java-enabled handsets (Anonymous, 2002, p. 23). A content market seems to exist in Japan. However, in the Netherlands, where the wireless carrier KPN has rolled out an i-mode service, the majority of wireless customers use prepaid plans for their wireless service. I-mode depends on subscriptions with monthly billing. Therefore, KPN must overcome two high hurdles: convincing user potential that they need a content service, and that they should abandon their prepaid plans.

2 The Tyranny of the Unk-Unks

It is difficult enough trying to settle on a business model when technologies are developing and morphing rapidly. There are countless uncertainties which need to be factored into plans and contingencies. But these are the simple unknowns. In these cases planners can at least anticipate and articulate what is unknown (e.g., "What happens to our model if reliable data compression jumps from the current x to z instead of y within five years?") But what is far scarier for potential players are the "unk-unks", "the unknown unknowns": those uncertainties that they didn't even know to consider in their evaluations (Compaine & McLaughlin, 1987).

What are the unk-unks in wireless? By definition if they could be identified they would not be unk-unks. A now-known surprise for the wireless industry was the blossoming of the 802.11 devices (Sandvig, 2004). There was nothing secret about the development of this protocol. But I have no doubt that the engineers and strategists at the wireless telephone enterprises did not give it a second's thought when developing their financial plans for 3G spectrum. However, almost coincident with the 3G auctions around the world, some entrepreneurs, experimenters, and free spirits were finding that they could provide many of the same data services that the telecom providers were hoping to sell using

3G. And with far less investment. Seemingly out of nowhere—the very insidious commonality of unk-unks—various "Freenets" and *ad hoc* networks of 802.11 systems rose to compete with the still developing 3G business model.

At this point it is not a certainty that 802.11 networks will in fact substantially compete with planned 3G services. But at the least it is the kind of surprise that can undo or undermine plans and strategies.

Another source of unk-unks is unexpected government regulation. For example, the accounting profession in the United States faced unanticipated scrutiny after 2002 when a series of headline events involving the financial statements of, among others, Enron, Worldcom, and Qwest. They created such an uproar that politicians were tripping over each other to introduce stern new laws and regulations.

Of course, perhaps the biggest unk-unks was the Internet. The Internet hit the popular conscious in 1994. That was when Netscape introduced the first widely available version of the graphical browser that turned the World Wide Web into a mass audience medium. There was a small community of academics and military contractors who had used the Internet since its inception in 1968. Yet in the wider world, there was virtually no recognition of the Internet as an alternative or a rival to the numerous attempts at developing a consumer online service. These began in the later 1970s with the British Post Office's Prestel. In the U.S, AT&T and the Knight Ridder newspaper group tried a system called Viewtron. The French government-owned telephone provider launched Minitel, a massive proprietary system. Back in the U.S., some text-based on-line services, such as the Source and CompuServe started in the 1980s. Prodigy and America Online began slugging it out for a graphics-intensive services in the early 1990s. As late as 1994, both AT&T, with its Interchange services, and Microsoft, with its launch of Microsoft Network, were still thinking in terms of a proprietary network (Compaine & Gomery, 2000, p. 438). All offered e-mail, but only for others subscribing to the same service. All of the smart people designing, funding and marketing these services missed the potential of the Internet which was right under their noses.

Thus, business models can be undone not only by the many known uncertainties by a developing communications technology, but may also be undone by the unexpected uncertainties that were not even on the radar screen.

3 The Media Model: Content, Process, and Format

While one cannot predict the models, one can provide some tools for analysis. To aid in conceptualizing the intersection of wireless technology and content it is useful to utilize an analytical model first presented in 1979, introducing a framework of *content, process* and *format* (Compaine, 1984).

Content. There are a multitude of ways in which we can express information content. Content may be data, knowledge, news, intelligence, or any number of other colloquial and specialized denotations and connotations that can be lumped under the general rubric of "information." Content is what fills up the papers in books, is captured on film, is sent over radio waves.

Process. This is the application of instruments, such as typewriters, computers, printing presses, the human brain, telephone wires, or delivery trucks to the creation, manipulation, storage, and transmission/distribution of content in some intermediate or final format. For example, a traditional newspaper relies on processes including entering thoughts of a reporter into a computer by manipulating a keyboard of a video display terminal with storage in the computer, and the eventual creation of a printing plate and distribution to consumers via trucks. Wireless is a another process—one option to getting bits or sine waves from point A to point B.

Format. "Print" or "audio" are essentially examples of formats in which some content can be displayed or otherwise manipulated by users. Words can come as speech or as squiggles. And those squiggles can be gouges carved in rock, toe marks in the sand, ink deposited on paper, or glowing phosphors on a screen.

The value of this framework is that it helps separate technologies from content and appliances from transport. "Television" is a good example. In its early years that term was used to refer to an appliance (the television set), to an industry, and to the medium that we came to know as images on a cathode ray tube integrated with sound from a speaker. When the only process for getting the content to the appliance was terrestrial broadcasting, then there was no real problem in using the term television to this entire chain: content (producer)-to broadcaster (process)-to video appliance (format).

Today, television is no longer a very accurate descriptor. The processes we use to watch and listen to the video box may involve terrestrial analog broadcasting or digital broadcasting. It may be delivered by a coaxial or fiber optic cable. It may be transmitted over microwave frequencies or by satellite. We can watch "television" from a DVD disk or videocassette. Or the process may involve a TCP/IP stream brought by what would have once been considered a simple telephone line. Thus, while content may not be all that different from 20 years ago and the appliance may still be called a television set, the processes have multiplied, with implications for regulators, content providers, and users.

One critical question which needs to be examined in determining content models for wireless is the value added by content to the newer wireless processes. If it remains speculative to declare what are workable models for wireless content providers, there are ways of strategically thinking about what might work. The top level guidepost is this: *How does mobility add value to content?*

While wireless network operators may be seeking new revenue streams by selling content, and content providers may hope for new markets by being able to sell to users on the move, there is also reason to ask when or whether third-party produced content will be a driver of wireless services. It has not until now.

There is no great epiphany in observing that mobility is very different from being tethered. Radios in automobiles, or carried to the beach, or taken on jogs, greatly changed how people used radio compared to the early days when the family gathered around the radio receiver in the living room to listen to specific shows, in much the same way television is still used today.

Up to now wireless devices have proliferated on essentially user-created content: primarily what we have to say to each other. In the forefront of testing content value-added, Japan's NTT DoCoMo's M-stage services let Japanese consumers receive content audio and video formats on their mobile phones. In its first year, it signed up barely 100,000 users. Contrast this to its launch of its person-to-person short message service i-Mode mobile Internet services three years previously, which needed only six months to sign up 1 million users (de Lussanet et al., 2002, p. 5).

There are, to be sure, opportunities for providers of content. But given the poor track record of selling content services over the Internet it may well be that the content will come from unexpected and nontraditional sources besides the vast pool of user-generated content. For example, many airlines with long distance flights these days are providing passengers with video screens of real time information on the position of the plane: distance from the start, to the destination, altitude, airspeed. Auto manufacturers have been looking at adding GPS and other "telematics" to their cars. These are example of wireless content, though they are not provided by traditional types of content providers.

Thus, wireless content may be provided by traditional content providers, such as publishers and producers, but may come from many other sources as well: real time data on position, *ad hoc* results from a data base search such as real time securities prices, talk and text, photos or video created and sent peer to peer.

Besides the many different forms of content, determining successful models is a function of local cultures as well. I have already noted that wireless itself comes in many flavors, each with its own technology and regulatory baggage: microwave, 3G, 802.11x, AM, FM, UHF, and VHF to name only a few.

4 Predicting is a Hazardous Occupation ... Especially When It Deals with the Future

There is a wonderful book called *Megamistakes*. It chronicles many famous modern projections and predictions and how they went wrong (Schnaars, 1989). It is useful to relive projections made by seemingly well-informed and authoritative sources:

- IBM's early 1950s estimate that the worldwide market for digital computers would be about 15.
- RCA's 1966 estimate of 220,000 computers in the world by 2000.
- At the other extreme, AT&T predicted the Picturephone market to have over 10 million users by 1980.

And if the nature of the projection is meant to help with a business decision, the devil could be in the details. There is the case of a market analysis company called Predicasts, which in about 1981 did a study of the market for home video playback appliances. In particular, they looked at the young videocassette and the nascent video disk segments, projecting the number of units likely to be sold annually for the ensuing five years. As I recall, their summary projection looked something like Figure 1. The top line is the total number of units to be sold each year (this is just representative here, not the original chart or its numbers). Looking back, the total number of video playing devices sold in 1986 was almost precisely what was projected in 1981. But Predicasts' analysis was that the dotted line, which represented video disk players, would outpace the lower solid line, representing video cassette recorders. In fact, video disk players turned out to be almost invisible through the 1980s, while VCRs became ubiquitous. Any manufacturer that made their plans based on the Predicasts model, or any programmer who struck deals for video disk licenses at the expense of the cassette format, would have missed an important market. So even when a prediction is right in the aggregate, it may be a disaster underneath.

Indeed, closer to our time, there are lessons to be learned about predictions of workable models from the Internet, particularly the dot.com bubble. The end of the 1990s was a period of the explosive growth of the Internet. It was a tremendously useful and productive, though short era. The initial dot.com bubble was right out of Mao: Let a 1000 flowers bloom.

Although much attention has been focused on the bust side of the era, quite a few of the 1000—or 10,000—sites found successful formulas. None, however, were automatically able to be duplicated. As noted previously, eBay showed that there could be a model based on owning nothing and selling nothing, but taking a small piece of the transaction. PayPal thrived by find-

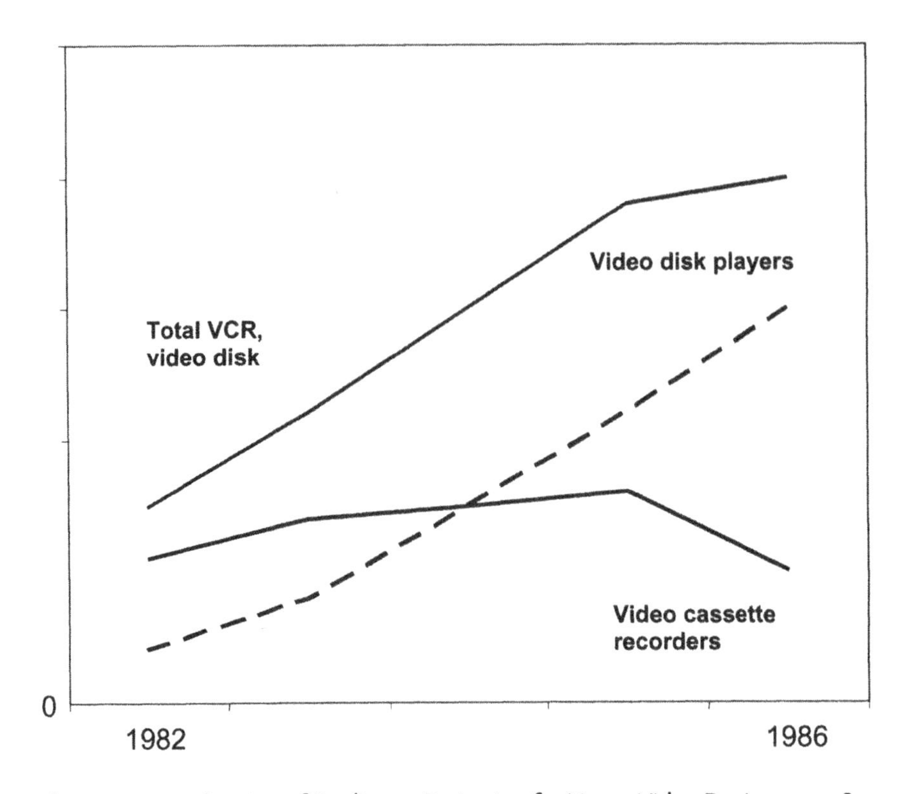

Figure 1: Generalization of Predicasts Projection for Home Video Devices, c. 1981

ing a way to serve the huge population using eBay. The Wall Street Journal Online provided evidence some hundreds of thousands of consumers would pay a significant sum for content—even content that they got as part of the newspaper many were already paying for.

Then there is Amazon, which found that there was a market for selling branded goods without having a physical retail presence anywhere. Travelocity, Expedia, Orbitz, and Priceline all found that consumers could do for themselves what they had been depending on travel agents to do for them—and save airline, hotels, and car rental agencies money in the process. And brokers learned that many investors would be quite happy to place their own trades—and many more of them—if they saved 90% of what it cost them to do the trade by speaking to a live order taker.

Some of these models may have been predictable. Booking for travel has been part of the mantra of expected online services since the days of videotext in the early 1980s. To the surprise of many prognosticators, on the other hand, the apparent success of *The Wall Street Journal* as a subscription model stands almost alone among mass audience news and information services. The advance predictors had expected that consumers would be far more forthcoming in providing a revenue stream for news than has proven to be the case. Also, both predictable and subsequently profitable have been the many sites offering content featuring overt sex. Porn was also an early driver of videocassette rental. Unseemly to some, but nonetheless a model that has proven robust across every format.

On the other hand I do not recall any predictions of an eBay-like service. Although some initial models suggested that transaction based income could be attractive, it was generally described in a context of payment services or consignments, not a matching service.

Another observation from the Internet as a lesson for wireless or other more nascent processes is that many of the breakthroughs were not initiated by the established players. Expedia, the first of the booking services, was created by Microsoft—not an airline or travel industry player. Priceline was also the creation of a nonindustry player. EBay was not a creation of Christie's auction house. The major online classified ad sites, Monster.com, and Hotjobs.com, did not emanate from either the newspaper business nor the executive search industry.

There is something of a pattern in this. Entrenched incumbents tend to fear that new processes will undermine their existing businesses and either ignore, try to bury, or circle the wagons when technologies open the door to a new type of process or format that threatens their content franchise. Newspapers, for example, were largely protected from new competitors in their markets due to the high start-up costs and reluctance of merchants to spread their advertising budgets over multiple newspapers. The Internet undermined that, especially when the most profitable revenue stream of the publishers—classified advertising—could be cherry-picked—while bypassing the cost of producing the editorial content that surrounded it in the print newspaper.

5 Is Content King?

We might expect that, in a time of proliferating processes for moving bits that content providers should be the king of the mountain. There is great logic there: Whether by a roll of celluloid or bits on a DVD, a theatrical "film" can be sold to an audience sitting in a movie theater or in their family room. A newspaper publisher houses a vast database of content, much of it updated hourly: report on a fire here, on a city council vote there. Sell it as ink smeared on paper or as bits transferred to a screen on my desk or a screen on my PDA—whatever works.

And yet it is not so simple. No newspaper ever went out of business for its lack of content. No Web site ever died for its inability to fill screens. Nor, as discussed previously, does content necessarily mean packaged by someone else, telephone conversations, VCR time shift (content someone wanted, but not now), fax, Internet instant messaging, and e-mail are all examples of content that is user created.

On one hand, not all content is created equal. The movie "Spider Man" grossed $114 million in the U.S. its first week of release. The second placed film only brought in $10 million. While content is important, it is not necessarily determinative of success. Good content, unique content, and the right combination of content with process and format are all factors that create value. "Spider Man" offered by a high speed wireless connection to the screen of a PDA might be worth less to most users than "Spider Man" on a large screen with stereo sound. On the other hand, the value of a stock price on that same PDA to a trader stuck in traffic may be worth the cost of a service that is several times the price of a movie ticket.

This gets back to my earlier questions: *How does mobility add value to content?*

Millions of consumers worldwide are paying more per month for a wireless telephone service than they pay for a wired telephone service. The content of both is virtually the same: chats with friends and parents, a child checking in, setting up a business meeting. It is the capability of doing that from the parking lot of the shopping mall, the traffic tie-up on the freeway, the park or the airport that adds value to the content. Traditional providers of content—the kind that someone might be expected to pay for—will have to learn how their content becomes worth something to someone on the move.

One recommendation in that direction is what Forrester Research calls "conversational content" (de Lussanet et al., 2002, p. 7). This is a label for the observation that user-generated content has been the driving force of wireless and much of the Internet. Drawing from the reality that customers have shown a greater willingness to pay for communication than for third party content, the generalized model for wireless is to seek opportunities for providing low value content with high value mobile communication.

In Japan, DoCoMo's struggling M-stage services only offer broadcast content that must compete with richer alternatives like TV, radio, and CDs. It does not enable any type of peer-to-peer communication at all. On the other hand, 4.4 million consumers use J-PHONE's sha-mail service, which allows them to take pictures with their mobile phones and send them to friends, family, colleagues, or customers—enabling them to communicate through content they create themselves (de Lussanet et al., 2002, p. 7).

Clearly, there are wireless models for content providers that work today. The notion of conversational content is as close to a formula that has been offered. It is consistent with the recent history of the Internet and communications services. And other than niche services such as financial data, the near term models for content are likely to fall into this realm.

As we live day by day through the change brought on by technology, we often may not be aware of what is different. Change is incremental, but may look awesome if we take a historical perspective. Much of what we take for granted today was talk in 1950s, in the labs in 1960 and 1970s, expensive, early innovator stuff in 1980s. A few years ago I dropped my daughter at the airport for a flight to visit her grandparents. My daughter, then 11, called me from a plane. "What's wrong?" I asked. "Nothing, just wanted to tell you we got off ok." The voice on the call was clear as if from a wired phone next door. Thinking of the marvel of my technology-challenged wife placing this call from her seat on a plane at 30,000 feet and 500 mph I replied, "Isn't this amazing?" "What is?" my daughter asked.

She had never flown on a plane without a telephone.

Endnotes

[1] For history of radio broadcasting in the U.S. see Christopher H. Sterling & John Michael Kittross, *Stay Tuned* (Lawrence Erlbaum Associates: Mahwah, NJ, 2002).

[2] Dow Jones reported that The Wall Street Journal Online had 646,000 paying subscribers in October, 2002. See *"The Wall Street Journal Online Wins Two Top Honors in 2002 WebArward Competition,"* Business Wire at *http://www.businesswire.com/webbox/bw.100302/22276385.htm,* Oct. 3, 2002.

References

Anonymous. (2002). Developing the Mobile Content Business, *Handheld and Wireless Solutions Journal,* Spring, 2002, http://intel.com/pca/developernetwork/solutionsjournal/spring_02/pdf/cybird.pdf.

Compaine, B. (1984). Content, Process and Format: A New Framework for the Media Industry. In B. Compaine (Ed.), *Understanding New Media* (pp. 69-95). Cambridge, MA: Ballinger Publishing Co.

Compaine, B. (1988). New Perspectives for Evaluating Media Competition. In B. Compaine (Ed.), *Issues in New Information Technology* (Chap. 3). Norwood, NJ: Ablex Publishing Corp.

Compaine, B., & McLaughlin, J. (1987). Management Information: Back to Basics, *Information Management Review, 2*(3), 15-24.
Compaine, B. & Gomery, G. (Eds.). (2000). *Who Owns the Media: Competition and Concentration in the Mass Media Industry,* 3rd ed. Mahwah, NJ: Lawrence Erlbaum Associates.
de Lussanet, M. et al. (2002). *Conversational Content Unlocks Revenue.* Amsterdam: Forrester Research.
Sandvig, C. (2004). An initial assessment of cooperative action in Wi-Fi networking. In *Telecommunications Policy, 28*(7-8), 579-602.
Schnaars, S. P. (1989). *Megamistakes: Forecasting and the Myth of Rapid Technological Change.* New York: Free Press.

6
Design Strategies for Future Wireless Content

John Kelly

1 Introduction

If you keep an eye out for the handiwork of mobile device and services marketers, especially in Europe and Asia, it is hard to miss visions of media pouring out of tiny gadgets. Bus stop posters, magazine ads, and signs in store windows show all sorts of exciting things happening on tiny screens: singers crooning, soccer players kicking, jazz artists trumpeting. Musical notes float out of mobile headsets, and the celebratory cry of "Gooooooaaaal...." meanders from a cell phone earpiece right across and off the edge of the page. To be sure, you also see images of happy people with phones to their ears, presumably talking to happy friends and family on the other side. But it is hard to escape the idea that if all you are doing with your mobile phone is making voice calls and (outside the U.S.) sending text messages, you can look forward to doing something much cooler with it.

This is of course precisely the idea this marketing strategy is trying to convey. The entire mobile telephony industry has a big stake in consumers opening their wallets to pay for a ride on the advanced mobile services bandwagon. To justify powerful new devices and networks with greatly enhanced data capacity, consumer adoption of high-bandwidth services must happen. And in the absence of clear precedent, the easy answer to the question "what will consumers want?" is "content."

So what is the of future of "wireless content?" Is the principal challenge we face how to get Paula Zahn's reportage or Britney Spears' vocal stylings onto our cell phones? Will we be reading the New York Times or Oprah's Book of the Month on our PDA's? Should we view mobile wireless devices principally as little, un-tethered TV sets, stereos, computer screens, and so on, in other words, a new channel for the distribution of content made for traditional and Internet platforms? Or, do we need to consider mobile devices (and the network behind them) as a new and different medium, with its own properties and potential?

I argue the latter, and in this article will address some of the key issues to understand in the design of new content applications for mobile devices.

After looking at key issues, we will consider some examples of advanced functions and services that might point the way to what the future holds in store. These are concepts, some in the lab, some in the imagination, some in trial deployments or limited commercial distribution, that embody the elements likely to come together as advanced mobile platforms mature.

But first, we must try to get underneath two dangerous assumptions common in the communications industries. The first is the old saw that "content is king," and the second is that high bandwidth is the necessary enabler of advanced services.

2 Challenging Assumptions

In his paper "Content is Not King," Andrew Odlyzko (2001) makes a compelling argument that while traditional content (i.e. professionally produced media) has a high profile and mind-share, it is dwarfed in revenues by connectivity (i.e., person-to-person communications). While content gets the glitz, one-to-one communications generate the real money required to support infrastructure on the scale of telecommunications networks.

And beyond revenues, connectivity also dwarfs content in terms of importance in people's lives. For example, if faced with a choice between giving up the web and giving up e-mail, people overwhelmingly report that they would give up the web (Katz & Aspden, 1997). And those who quote statistics to show that web data outweighs e-mail data on the Internet should consider carefully the difference in "attention value" between one bit of personal e-mail (expensive) and one bit of a banner ad graphic (cheap).

For our purposes, the main point here is that people are much more willing to pay for connectivity than for media content. For the most part, content is not something consumers buy. Quite the opposite, it is something advertisers use to buy access to consumers' attention. And if you consider the types of content that consumers do buy, little of it is the sort of thing that has any real prospect of a wireless play. People generally want to watch movies on the biggest TV sets they can afford or accomodate; eBooks are not doing well; and despite some industry excitement about the idea, individualized wireless downloading of high-quality music is unlikely to be feasible for a variety of cost, efficiency, and usability reasons. And as for advertiser-supported media, interruption based ads upset consumers, who consider their mobile devices to be very private, personal accoutrements. (There are good opportunities for contextual advertising, as will be discussed later, but only on an opt-in model that is designed around consumer needs and not likely to be tied to a media play.)

This argument stands the strategy of many wireless operators on its head. These operators see wireless content as a high-margin business that will justify the expense of deploying the advanced, higher-bandwidth networks required for multimedia. They think that consumer demand for media content will save them from being in the uninteresting and presumably less profitable commodity business of providing "dumb pipes."

There is a certain circular logic here. To build advanced networks, one needs the higher profits (supposedly) available from being in the less heavily regulated content business. And to deliver content that consumers will pay for (directly, or indirectly through advertising), one needs high-bandwidth networks. But the failure of the Internet as a "content" medium should be a lesson to those thinking about wireless in similar terms.

If we consider the landscape of wireless platforms' capabilities in two dimensions, bandwidth and "social scope", it is easy to see how lower bandwidth services not only form the core of the revenue base, but have the strongest still untapped potential for advanced mobile data services. In terms of bandwidth, current systems easily handle voice, of course, plus data services that can be accomplished with even less bandwidth. EMS and MMS do not require vastly more bandwidth than voice EMS could use less. 2.5G would be good enough for these services. Most communications can be accomplished with low data rates, and those that cannot, like broadband quality video, really need the much higher bandwidth of WiFi (also called 802.11, or WLAN) to be feasible.

In terms of "social scope" of the communication, we can distinguish personal, one-to-one communications on one extreme—like phone calls—and one to many, mass communications on the other. In the middle are the increasingly robust one to some communications typified in the old days by community and interest specific newsletters, mailing lists, etc., and exemplified today by online chat, online interest groups, listservs, family, community and interest-specific websites, and so on. It is important to note that even as online media plays are dying off, these one-to-some "community" features are thriving on the Internet, and could prove successful in the wireless domain as well.

If we accept Odlyzko's argument, that the traffic and revenues needed to support a communications infrastructure can only come from the more personal communications, we should recognize that these are also the ones most easily accomplished at lower bandwidths. In the traditional, one-to-many domain of "content" the only current examples cited as wireless success stories are screen graphics, ring tones, and wireless alerts. Of these, we would argue that only wireless alerts actually function like traditional content. Screen graphics and ring tones are more about saying "check this out" to your friends, a "social application" more than a "content" one.

So a question for operators is whether there will be sufficient consumer demand for media-rich, one-to-many applications to justify the enormous cost building high-capacity networks, not to mention the time and money spent trying to build a media business on top of a communications business. Unfortunately for mobile operators, 3G is too limited for truly compelling media applications, and overkill for basic services known to be useful and in-demand. It would be better to move beyond the idea of mobile devices as media players.

The objection that the future may bring applications we cannot foresee has the virtue of always being true. But it is weakened by two considerations. The first is that the potential for compelling low data-rate mobile applications has hardly been tapped. There is a long way to go in exploiting the value of the bits we can send at 9.6 kbps, and no reason to think that those bits will become more rather than less valuable when they can zoom around at 3G speeds. Using Java and other ways of making the ends of the network (mobile device and back-end) more intelligent, even very low data rates can generate compelling, rich experiences.

The second, related, consideration is the rich history of the mistaken belief that if people like something now, they will like it a lot more if you "improve" it—even if they are not asking for the improvements. HDTV, Quadraphonic sound, and the video-phone (particularly relevant here) all leap to mind. This is not to say that there aren't "killer apps" out there to drive adoption of mobile data, just that adapting mass media for wireless distribution is probably not one of them.

None of this should be taken to mean that today's 2G systems are perfectly adequate and need no improvement. But among the technical advantages of 3G networks, the "always-on" feature of packet-switched systems is significantly more important than the higher bandwidth. In this sense, "2.5G" networks (e.g., GPRS) may be good enough.

There is, however, an analogy between the unforeseen benefits that could lurk in the deployment of advanced mobile networks and the real benefits of wired broadband over dial-up. Despite marketing that emphasizes rich multimedia content, broadband adoption in the U.S. has principally been driven by consumer frustration with slow page-loads and a desire to save time (Rappoport, Kridel & Taylor, 2002). In other words, for the most part people pay for broadband to better enjoy principally text based web pages, not video, animation, or other rich media. Similarly, high speed wireless networks could succeed mainly by making low data rate services already available on current networks function better, and perhaps less expensively. The latter point is significant, since with the exception of SMS, the high cost of even very low datarate services is a major inhibition to consumer adoption in Europe (Jupiter MMXI, 2002).

3 Designing Wireless Applications

So how should we think about the future of wireless content? The key is to figure out what the special properties of the platforms are, in other words to focus on what is unique about networked mobile devices rather than how they can act like movable, miniature versions of things we already have. Designers of wireless applications face numerous challenges grappling with this problem. Mobile devices have severe restrictions imposed on form factor, input and display, battery life, processor speed, communications range, data rate, and memory, to name the most obvious limitations. And then there is the big mystery of what people will actually want to do with them.

It is very hard to understand what consumers will want by brainstorming in a vacuum, and consequently it is critical to track, and attempt to leverage, the practices and cultures rapidly evolving around digital technologies. It is also important to understand the key properties of wireless mobile devices, how consumers adapt to them, and the implications of particular business models on both user experience and content development.

Taking a look at the major dimensions, we will now illustrate key opportunities and principles.

3.1 "Anytime, Anywhere" vs. "Here, Now"

In terms of performance and features, mobile devices are inferior to desktop computers in nearly every respect: slower processors, smaller displays, restricted input methods, tiny amounts of memory. As networked communications devices, they offer nothing in terms of features over the PC and phone, if you consider that Instant Messenger, ICQ, etc. are the equivalent of SMS on the PC, and that in Europe at least, carriers are rolling out wired phones that implement SMS for the home or office (Anonymous, 2002). And they are more expensive generally than landlines for voice and dial-up for data—and even broadband data in many cases.

This is all to drive home the fairly obvious point that the only advantage mobile devices have is that they are in fact mobile. They accompany their user everywhere he or she goes, providing ubiquitous personal connectivity. "Anytime, anywhere," as the slogan says.

The "anytime, anywhere" concept is the first dimension of mobile applications, and refers to the idea of extending communications services available on other platforms to mobile ones. It provides social connectivity like SMS, e-mail, and now Instant Messenger, as well as group messaging like that offered by the mobile community enabler Upoc (a service supporting wireless messaging groups, such as New York's popular "nyc celeb sightings"). It also entails things like wireless alerts, news, sports, weather, bank account updates, stock price changes, and so forth. It essentially refers to an extension of cyberspace into the mobile dimension, allowing services you can enjoy by other means to be more platform and device independent. Most mobile services available now fall into this category.

The second dimension of mobile applications is commonly referred to as "location-based services," and means functions that are specifically about "here" and "now," quite the opposite of "anytime, anywhere." But the term "location-based services" is too narrow to capture the implications of devices becoming much more intelligent, about not just where they (and their users) are, but what else is around them, what their users are doing, and other aspects of their proximate contexts. The term "Hypernet" better captures the idea of a vast diffusion of networked devices—mobile, embedded, implanted, and so on, working together to map the power of the Internet onto the real world.[1] One easy way to think about the mobile device in this way is as a "remote control" for life, like the one we all know and love from the couch, but one that comes out and about with us wherever we go.

We probably get a better taste of what the Hypernet will be like by looking at GPS devices, and imagining how far you could go with RFID-based (Radio Frequency Identification, passively powered data chips that can be very small and embedded in just about anything) systems like EZ-Pass (an automated toll collection system), than by thinking about pop-up coupons for the nearby Starbuck's. Consider the Rhino[2], a two-way radio/GPS hybrid from Garmin, that brings features available to the military to the public. As you and your friends tune to the same channel, the radios not only transmit your voices, but also continuously share everyone's coordinates. You can see yourself and your friends on a map as you hike, ski, shop, sightsee, and so forth. Implement the concept on mobile phones and it would be great for families at the mall or amusement park, not to mention giving directions, improving travel, or any number of other things. Clearly, as with E-911, there are technical and privacy issues to work out, as well as important issues of location granularity, but in the end there are a number of ways for devices to know where they are within margins of error tolerable for a host of compelling applications.

Beyond knowing where people are, the Hypernet includes the idea of knowing about objects in the environment. Behind automatic tollbooth technology like EZ-Pass and some anti-theft tags on retail items are RFID's. Read from short distances by radio waves, RFIDs work like next-generation bar codes. They can identify cars, books, clothing, art hanging on a museum wall, or virtually any solid object. How often do you see someone reading a book and think you would like to have that book too? Now, you need to ask or lean in to read the title and author, remember or write down the information, go to the bookstore or find it online. What if you could just point your phone at it, click "order," and just wait for it to show up at your door? While browsing DVD titles, RFIDs in the DVD case could trigger wireless playback of the film's trailer. Artwork in a museum (or anywhere) could trigger delivery of background information. The concept of the Hypernet takes the power of cyberspace—the abstract, location-less Internet and brings it into the here and now, able to know where you are and what you are doing, see anything that can be made machine-readable, and control anything that can be made machine-controllable. It will take a while to get there, but it is a more compelling vision to build design concepts and business ideas around than "location-based services."

Another critical implication of both the "anytime, anywhere" and "here, now" aspects of mobile applications is that mobile devices will not simply be valuable as end points on a communications network. Of course consumers will want to make phone calls and read their news alerts, but in their "remote control" mode devices can serve as the 'connective tissue' between a host of functions, machines, and transactions in daily life. Mobile devices are likely to grow in this capacity as other systems become increasingly tied together via the Internet. Just as users of ReplayTV's latest digital video recorders can program them from a web browser, a mobile user could hear a song on the radio and use her phone to order it downloaded to her PC or entertainment center at home. Of course most of this is still pretty far out, but the principle is one to remember. Consider that a number of successful SMS-based m-commerce models in Europe work well precisely because they view the phone simply as one link in a chain that ties the physical store, the Internet, and mobile interactions together as a single unified experience for the customer (Klym, 2001).

3.2 Understanding the Mobile User

A complaint about WAP heard in developer circles is that creating applications is too difficult and requires even an experienced web author to learn a whole new bag of tricks. By contrast, creating i-mode sites is easy for anyone with the kind of basic HTML skills any web developer will already have. Part of the problem is that WAP sites must be engineered to work with a bewildering array of non standard devices, with different screen sizes, input methods, and specific capabilities. I-mode, by contrast, has the luxury of working with devices designed specifically for its service. Another part of the problem is that WAP was overengineered to allow the kind of complex interactions possible on the Internet. Whereas i-mode uses cHTML, a simple, stripped down version of HTML, WAP requires WML, a more feature-rich but complicated standard. WAP ignored a cardinal rule of wireless interaction design: Simplicity is king.

Wireless interactions and Internet interactions are very different, and not only because wireless devices have small screens and keypads. The user contexts are different. When using a PC, one is generally at a desk, is expecting to spend several minutes if not several hours in the same spot, and in any case—and this is the important part is focused on the screen as the primary object of attention for a prolonged period. The PC is normally the principal interface of one's primary dedicated activity. This fits nicely with the fact that it has a large screen and good input methods, but it needs to be understood independent of that fact—from the perspective of the user's expectations, purposes, and contexts.

Wireless interactions generally occur in a very different context, one in which the user is out and about and has the world itself as the principal interface of their primary dedicated activity. To make a PC metaphor out of it, the real world is your screen, your hands and feet are your mouse and keyboard, and the wireless device is a tiny little pop-up window that distracts you from time to time or allows you to perform some simple little function off to the side of your main activity. It may sometimes be assisting you with that activity, sometimes interrupting to alert you to something else you need to be aware of, sometimes allowing you to take care of something quickly so you can get back to your main activity. In most cases though, the mobile device is not "the show," it is more like the beeper that goes off while you are watching the show. Mobile interactions for the most part are distracting or assisting interactions.

This means that designing good wireless applications is not a function of miniaturizing web-style interactive experiences. Rather it is about simplicity, usefulness, and clear limits on required attention. The "mobile attention span" is very short. With m-commerce for instance, complexity is something better built with multiple quick interactions spread over time than with deeper menus or more robust forms (Klym, 2001). In terms of data types it gives text and audio higher value than video. Text is good because it is "on your time," meaning you can read a chunk, look away, and come back to it, usually without missing anything. With audio, like any time based medium, you are "on its time," but at least your visual attention is free to stay with the world around you—one reason radio is so popular with people driving or working with their hands. Video is the most "attention-expensive" medium because you are on its time and you cannot interleave your attention to it with any other activity. This is not to say that more attention-expensive experiences will have no place on mobile platforms, just that users will engage with them only in a very limited range of circumstances such as the daily train commute that has helped certain i-mode services in Japan. Even entertainment experiences like gaming will benefit by engaging players not just with the device, but with the world around them.

3.3 Learning from i-Mode, European, and U.S. Adoption

The popularity of particular services, the willingness of consumers to pay for them, and the best way to design and market applications will depend on a host of local cultural and market factors. For instance, Americans might get to the point where customized phone face plates are as big a deal as they are in Europe, where every mobile phone store has a plethora of colorful, branded alternatives, and street vendors and trendy back-street shops in major cities like Paris and Budapest sell hundreds of unauthorized custom designs. But they are unlikely to ever go to the extremes of customization and techno-fetishism found among teens in Japan. Similarly, Americans will probably (eventually) adopt texting because of the communicative efficiencies that have led to its success elsewhere, but will not share the added incentive that Japan's high social barriers to verbal and face-to-face communication (often seen as "shyness") give the Japanese. And there are significant user context differences between societies. For instance, Americans tend to spend more time in their large homes (and with their large TVs), while Europeans and Japanese spend more time in restaurants, cafes, and other public spaces. And while many Japanese spend their daily commutes crowded onto trains and needing something like i-mode (or a good book) to stay occupied, Americans sit in cars and listen to the radio.

Looking broadly at international mobile behavior and adoption, as well as online behavior, there are some key themes with implications for the future of content and services. One is what could be called the "social power" of mobile communications. Another covers a bundle of key issues in the overall usability of services, including a look at the very difficult issue of "walled gardens" versus open environments.

4 Dimensions for Design

4.1 Dimension 1: The Social Power of Mobile Communications

Because mobile devices are such personal objects and typically accompany their users everywhere, they tend to be seen as extensions of self, much like cars or clothing. This is seen most obviously in the case of custom phone faceplates. But the phenomenon extends to custom ring tones and screen animations. While the latter are sometimes described as "content," as though downloading a ring tone would be the same type of consumption experience as downloading a favorite song for your MP3 collection, the value of ring tones and animations is not really in one's personal enjoyment of them. Rather the value derives from other people experiencing them, allowing the phone's owner another avenue of expressing social identity.

Similarly, services or content that play a role as "social currency" have become widely popular, especially in youth culture. In London, a popular youth behavior is sending around little text animations, much like the "ASCII art" that adorned many people's e-mail signatures before embedding custom pictures became a possible (and preferred) means of marking one's identity. Most of the animations are comical visual gags, and a great many are pornographic. The value of these animations is in having them to share and trade, and especially in being able to show them around one's table at the pub. Like having a great joke to tell (or for kids, a valuable Pokemon card to trade), these bits of mobile content fit seamlessly into natural social life. Another great example is big-screen public SMS displays, something gaining popularity in parts of Europe. Youth venues (bars, clubs, raves, concerts, etc.) feature large screens and special phone numbers for the public display of SMS messages, which range from the standard "Jorg luvs Elena" sort of thing, to much more inventive and insider-oriented social ploys. This is a phenomenon that probably stands to gain a lot from the widespread rollout of MMS services.

Another area where we see the social power of mobile communications services is in the emergence of mobile group applications, including games and interest groups. An intriguing example from Sweden is a game called "BotFighters," the first commercially implemented Hypernet game played over mobile phones.[3] (One could argue that Electronic Arts' "Majestic" has the same claim, but mobile phones were just one of several channels available for playing it, and not an essential component.[4]) In BotFighters, the players set up a "bot" using the web, customizing such elements as weaponry, shields, and special capabilities, and then "battle" one another in and around Stockholm using mobile phones as game controllers. The "game board" is the environment itself, since the network can track player location. Battles require physical proximity. "Health packs" and "ammo" are "picked up" on street corners, in parks and other locations by physically walking, running, or driving over their "location." The game is persistent (meaning the world of the game continued around the clock), and it developed a dedicated group of players willing to jump out of bed at 3 a.m. and drive half-an-hour to join battles.

Some of these "one-to-some" applications, as you would expect, mirror popular Internet applications like dating sites. One of the most popular categories of "unofficial" i-mode sites is for singles "hook-ups." In the U.S., Upoc has introduced mobile group communications services, some of which are building successful ongoing user communities. Upoc allows users to join groups which function essentially like mobile messaging listservs. Users can join public groups, some of which are sponsored, and form private groups for sets of friends or colleagues. In practice, these groups are similar to communities of interest that develop on the web, but support much more location-oriented activities, like arranging nightlife activities. A particularly popular one in Manhattan is for celebrity sightings. Upoc and BotFighters are two examples of how "content" for mobile platforms can flourish in the space between one-to-one communications and traditional mass media.

4.2 Dimension 2: Macro Issues of Usability

Normally, "usability" refers to things like clarity of visual interface, logical navigation, ease of input, the number of clicks or menus one must go through to achieve a desired end. But when looking at the wireless landscape we need to consider what we could call "macro" issues of usability that have dramatic effects on consumer adoption. These are things that frame the entire process of getting to, or deciding to use, any particular interactive service. And here, i-mode has some valuable lessons to teach us about designing a wireless business that puts the user's interest, rather than a defensive business model, first. This is an area where U.S. and European players have failed.

Consider how one accesses services on i-mode. There are two type of i-mode sites, official and unofficial. Official sites are selected a few at a time from among hundreds of applications submitted by third party players.

DoCoMo's priorities in selecting sites are to insure quality and a robust mix of services for their subscribers. Beyond the very early stages of setting up an initial collection of services to launch i-mode, DoCoMo has not attempted to own or develop its own content offerings, or to take an interest in any particular official site at the expense of another. Menu placement on i-mode is like a constantly updated Top-40 list, with the most popular sites floating to the top. DoCoMo opens its billing system to official sites, so that subscriptions and services show up together on the customer's mobile phone bill. However, DoCoMo makes 70% of their revenues on the data traffic generated by i-mode (Lemon, 2001), and takes only a 9% cut of revenues billed through their system. Unofficial sites, which can be created by anybody, are easy to access and bookmark with methods that let the user avoid typing in long URLs on a numeric keypad (as one typically must do with non-featured WAP sites). Unofficial sites do not share DoCoMo's unified billing system, but there are popular, easy-to-use third party billing systems available for many popular unofficial sites.

U.S. and European operators have taken the opposite tack, and consumers have stayed away in droves. By and large they have each created their own walled gardens, with very high walls. Menu placement is on the basis of payment or strategic partnership, not user interest. Unified billing for online content and services has not been successfully implemented, leaving consumers with the need to establish individual billing relationships with vendors and service providers, or to use third party Internet payment systems that are still struggling to achieve mass adoption. U.S. and European operators also seek to extract much larger percentages of revenues from their content and service providers. The key lesson of i-mode that should migrate westward is that to build traffic, the user's interests must be put first (Funk, 2004). (It should be clear as well that 9% of a lot is worth more than 40% of nothing.) An analogy can be made to the different approaches in the U.S. and Europe to mobile competition. One can argue that to maximize consumer interest it is better to standardize platforms, and compete on services.

Another macro-usability issue is how the usability factors on the micro-level can create larger implications for wireless content and services. We have touched on some of this already: Consumers like video, but they seem to like it best on their TV sets. To consumers, the value of broadband connections is not that they deliver rich media; rather, broadband is valuable because the consumers time and attention are not wasted in the process of first connecting and then using the web to do the same things dial-up web users do (namely access text and static graphics). Also, the young will adapt to new interfaces more easily than older folks. All of these things give us parameters for creating appropriate new content.

But also look at the experience of financial services firms, many of who rushed into the wireless space. What some have discovered is that although simple wireless information services (like custom stock alerts) are popular, customers do not want to trade wirelessly. They prefer to do that on a PC with its more robust tools to analyze information and control transactions. The lesson is that unified services that span platforms, and accurately capture the strengths of each platform, are positioned to create total user experiences that consumers will find valuable. And since it may be years before revenue models arise to support pure wireless plays, this offers another advantage to firms that can interface with their consumers across multiple channels.

5 Wireless Application Concepts

Knowing the broad guidelines and parameters for wireless applications, it is nevertheless challenging to envision the kinds of things people will find useful, let alone to predict "killer apps." Initial predictions of future usefulness for the printing press, telegraph, phonograph, telephone, television, computer, and Internet often proved wrong. So in the spirit of playful speculation, here are some concepts for the future of wireless that at least try to incorporate some of the principles we have discussed. They are certainly not meant to be predictions, simply illustrations. Nor are they original ideas, but rather pulled or pieced together from prototypes, test deployments, design demos, ideas for linking existing products and services, discussions with colleagues, and so on.

5.1 Concept 1: Wireless Digital Photography

The largest amount of personal data in a typical American family's home is often its photo collection.[5] Not surprisingly, of all the various digital gadgets available on the market these days, the heavyweight champion of consumer adoption is the digital camera. On the PC front, Apple's consumer oriented iPhoto provides consumer-friendly home digital image archive management. And Sony and Ericsson have teamed up to build wireless devices, including an MMS-enabled mobile phone with detachable digital camera. This builds on what are already a host of camera attachments for PDA platforms, also equipped with wireless connectivity. Wireless digital photography is already here, and the opportunity exists to put all these pieces together into services for seamless multi-platform digital photo management.

5.2 Concept 2: Voice = Value

A mobile phone and a digital voice recorder already share a lot of the same components, and they could be merged into a single device without negatively impacting device size and power consumption. A number of voice recording add-ons are already available for PDAs. The most advanced digital voice recorders have very robust PC connection features that allow for voice file data-basing and automated transcription. Mobile phones could wirelessly link up to similar services from anywhere. Taking advantage of server-side processing, a host of advanced management, transmission, transcription, and translation services are possible. The inclusion of significant memory for audio storage in the device could allow an attractive business in voice-quality audio subscriptions (which would require much less data than music-quality audio files). "Personalized radio shows," audio books, language instruction, and a host of other voice content could easily be available for faster-than-real-time download, seamlessly billed through the mobile operator.

5.3 Concept 3: Picture-to-Text, with Translation

The ability of a wireless device, limited in its onboard processing power, to take advantage of server side processing leads to other interesting possibilities as well. A project out of IBM's Almaden Research Center[6] demonstrates how digital photography, wireless transmission, optical character recognition, and automated translation can work together as a seamless application. A traveler sees a puzzling sign in a foreign language and snaps a picture with their mobile device. The picture is transmitted wirelessly to a service that automatically analyzes the photo for text, runs optical character recognition, translates it into her native tongue, and sends the text back to the device. She is able to look at the screen and see the translated text superimposed over the image. For travelers this could be a common, high-volume tool, if priced appropriately.

5.4 Concept 4: Rich Locations

Because of data-rate and reliability issues that give it a lower quality of service than wired broadband options, 3G is not really capable of delivering high-quality rich media. WiFi is capable of it, given its 10-Base-T Ethernet-level speed, but has a more restricted range. This makes it well-suited as a method of delivering rich media and data services related to a particular location. In a pilot project, visitors to Disney World can get specialized Compaq iPaqs that know their location and provide simultaneous audio translations of characters' live performances around the Magic Kingdom—in French, German, Japanese, Portuguese and Spanish—as well as information for rides and shows.[7] A text version is available for the hearing impaired. The possibilities for enhancing museums, monuments and historic locations are considerable.

5.5 Concept 5: The "Lovegety" meets Online Dating

The "Lovegety" is a small wireless gadget from Japan that fits on a key chain and is named for its ability to help its owner "get love." The original and its imitators never caught on in the U.S., but are widely available in Asia. In Hypernet terms, it offers a proximity-based service that alerts two suitable singles when they pass within a certain distance of each other, say at a bar or nightclub. An owner inputs a simple personal profile as well as the desired criteria of the other party, and when it encounters another Lovegety broadcasting the right stuff, the devices talk to each other, and then beep or vibrate to alert their owners that they should do the same.

The profiles stored in these devices are rudimentary. But the opt-in proximity alert concept, melded with the more sophisticated profiling and matching capability of today's advanced (and popular) online dating sites, could prove interesting to singles if implemented in mobile phones.

5.6 Concept 6: Persistent Wireless Gaming

As we discussed with Sweden's BotFighters, this one has been done. But it hasn't been done on a very large scale, with sponsored awards, in conjunction with major youth events like music festivals, or in very high-density urban environments like New York or Tokyo. There are a number of ways to take the persistent wireless gaming concept further than BotFighters. Persistent gaming attracts very large, fanatically dedicated players. It also generates a lot of money. According to some reports the virtual world of the largest persistent game, EverQuest, actually generates a per capita GDP of $2,226—not bad for an imaginary place.[8] As with Electronic Arts' Majestic, which failed gloriously due to problems in the games design, persistent games can be accessed via different channels, for instance tying web, e-mail, landline phone, fax, and mobile devices together as a seamless experience, and offering opportunities for cross-channel partnerships and sponsorships.

6 Final Thoughts on Design Strategies

The future "content" of wireless devices will be unlike traditional media content, rather it will consist of applications that take advantage of mobile devices' unique properties. Mobile applications will be able to leverage two different dimensions of usefulness: the ability to extend access to the kinds of services available via the Internet (extension of cyberspace; "Anytime, Anywhere"), and the ability to harness the power of networked intelligence to assist with life out and about in the real world (Hypernet; "Here, Now"). Successful future services are likely to reflect several key aspects.

- *A focus on user-centered design:* Firms must find ways to make money, but users will not be shoehorned into business plans. Success will be more likely when the user's needs are put first, and design is based on a detailed understanding of user contexts. Important factors include the shortness of "the mobile attention span," and the ability to build complexity of interaction over time rather than with complicated interfaces.
- *Communications, community, and information functions:* Content is not king on mobile devices, if we understand "content" to mean professionally prepared media for consumption by an "audience." One-to-one communications are likely to dominate traffic and revenue, and information services (rather than entertainment, although perhaps *about* entertainment) may be popular as well. There are opportunities in the space between these things, where one-to-some (or group) applications can expand their online popularity into the wireless space.
- *From "location-based services" to the "Hypernet:"* Taken together, the ability to pinpoint user location and the ability to "tag" the physical world allow a range of new services that can augment users' abilities in the world at large. A common metaphor is the mobile device as a "remote control for life."
- *Understanding evolving wireless culture:* Interesting cultures and social practices are evolving around wireless devices, just as they have evolved (and continue to evolve) around the Internet. Understanding this evolution is valuable for designing new applications.
- *Meta-narrative and gaming as foci of wireless entertainment:* While entertainment will not be the primary use of mobile devices, it may have a role. Traditional content providers can explore the development of "meta-narrative" applications, much like websites that provide extra character information and story details for an ongoing dramatic series. The paradigm is the extraordinary meta-narrative activity around *Star Trek*. Gaming also shows strong promise, including Hypernet concepts and persistent gaming.

It is likely to be a slow build toward a future of wireless content in which technological capabilities and consumer needs find a good marriage. A number of hurdles must be overcome, but perhaps above all else the telecommunications industry (carriers and equipment vendors alike) needs to figure out how to promote innovation. Arguably, like the music industry in the case of online distribution, it is stifling it, due to rival standards. The creation of effective standards would allow seamless delivery of third-party services across carriers and would likely spur innovation and consumer adoption. Until then, things will move more slowly in the U.S. than in key European and Asian markets.

Endnotes

1. Analysts Digital 4Sight have an excellent series of reports entitled the "Hypernet Revolution," which cover a range of key issues in advanced mobile services.
2. See: http://www.garmin.com/products/rino/.
3. For an excellent discussion of mobile games, including *BotFighters,* see: Lauf, R. & Cosgrave, D. (2001). Mobile Games: Play on the Go. *Digital 4Sight.*
4. *For a good description of Majestic, see:* http://www.wired.com/news/business/0,1367,43944,00.html.
5. See Lyman, P. & Varian, V., How much information, http://www.sims.berkeley.edu/research/projects/how-much-info/index.html.
6. See: http://www.technologyreview.com/articles/prototype41201.asp.
7. See: http://news.com.com/2100-1033-802775.html.
8. See: http://www.newscientist.com/news/news.jsp?id=ns99991847.

References

Anonymous. (2002). More Power to the Thumb. *The Economist Technology Quarterly,* June 22, p. 11.

Funk, J. (2004). *Mobile disruption. The technologies and applications driving the mobile Internet.* New York: John Wiley & Sons.

Jupiter MMXI. (2002). European Mobile Internet: European Market to Take Off in 2003.

Katz, J. E., & Aspden, P. (1997). A Nation of strangers? *Communications of the ACM, 40*(12), 81-86.

Klym, N. (2001). The Hypernet-Enabled Customer: Designing Wireless Transactions. *Digital 4Sight.*

Odlyzko, A. M. (2001). Content is Not King. *First Monday, 6*(2).

Rappoport, P. N., Kridel, D. J., & Taylor, L. D. (2002). The Demand for Broadband: Access, Content and the Value of Time. http://www.sess.smu.edu.sg/newsletter/pdf/Rappoport_1.pdf.

7
Mobile News Design and Delivery

John V. Pavlik & Shawn McIntosh

1 A Brief History of Early Wireless Media Content

Since radio pioneer Lee de Forest first delivered audio news reports and music via his wireless arc-transmitter in 1916 (Anonymous, 1916), wireless delivery of media content has been an increasingly important part of the mass communication landscape in the U.S. and around the world.

As much a self-promoter as a radio pioneer, de Forest observed, "Personally I can see no reason why the wireless telephone transmission of news in the near future will not be a regular means of communication, and a very valuable one, too, in supplementing by bulletin the various editions issued by the metropolitan newspapers. All that is needed is the news, and a comparatively few, well-located, high-power stations capable of covering the entire country. Already we have in the United States, I should say, at least 200,000 amateur wireless outfits waiting to receive news and music by the wireless telephone."

De Forest realized that wireless transmission of news was economically viable as well, noting that wireless news publishers would be the only ones in the world "who are not affected in one way or another by the higher cost of paper." Since the experiments of 1916, the 20th-century witnessed dramatic growth in wireless media, both in their content and distribution. Audiences have swelled to the billions around the world. The technology for mass media has changed significantly as well, including, the development of FM radio, wireless video transmission, satellite transmission, and the invention of mobile audio and video receivers. With the invention of the transistor and then the transistor radio, listeners were able to tune into audio media programs from their cars or portable radios.

Yet wireless media content itself has evolved relatively little beyond becoming somewhat more sophisticated in style, presentation, or production value. The basic model of producing audio, and later video, programs, either live or recorded, for one-way distribution to a listening or viewing audience has changed little in the eight decades since de Forest's time. The same content is often distributed to audiences via wireline media as well as to fixed-position devices. The content is generally in no way designed for mobile media, and is virtually identical whether received on a fixed-position or mobile device. Radio content, since it is in audio format, is the only type tailored for mobile or mobile wireless devices. There are some services, such as Vindigo (a subscription service that allows users to put phone books, directory services, guide books, and maps on mobile devices such as the Palm), that provide marketplace information such as restaurants, shopping or services (e.g., the nearest ATM) relevant to a user's location or destination.

The development and convergence of a series of wireless and mobile communications and computing technologies in recent years, however, presents a rare opportunity to dramatically transform media content for mobile wireless communications, especially in the design and delivery of news and information (Pavlik & McIntosh, 2003). News organizations are beginning to explore the implications of these technologies for news design, production and delivery. News services, such as the Associated Press (AP) and Reuters, are especially interested in mobile wireless technologies both for their reporting staff and audiences because of the speed and flexibility these tools provide. On one occasion in 2001, AP and Reuters both used mobile wireless technologies to report a much-anticipated decision in a court case in Lower Manhattan. AP reported the verdict by mobile phone. "We then got additional color material via Blackberry® from the courtroom," notes the AP's Rick Spratling.[1]

Using a computerized system that monitors the newswires, Reuters discovered that AP moved the story on the wire first by a few seconds—a critical span for traders.

Today, mobile wireless technologies are perhaps most important during times of crisis, such as during the immediate aftermath of the Sept. 11, 2001, attack on the World Trade Center when many telephone land lines were damaged or overloaded in Manhattan.

Against this backdrop, this chapter explores experiments with new forms of media content delivered to or accessed by consumers equipped with mobile wireless communications, conducted by a research program at Columbia University.

2 Today's Technology and Trials: The Situated Documentary

Through an interdisciplinary collaboration between the Center for New Media in the Graduate School of Journalism and the Computer Graphics and User Interfaces Laboratory in the Department of Computer Science, we have been developing and testing a new form of news content for mobile wireless communications. We call it the "situated documentary." The name derives from the fact that the audience member experiences a documentary when at the location where the events reported on originally occurred. The documentary is thus "situated" within the context of where it took place.

The situated documentary uses a combination of mobile and wireless technologies assembled into a "mobile journalist's workstation," or "MJW."[2] In order for the situated documentary to work properly, it must not only be able to track where the user is standing but must "know" which way they are looking in order to superimpose the correct scenes over what is actually being looked at through a head-worn display. The computer must be fast in order to process large amounts of information that includes video clips and other multimedia. The MJW must be interactive as well, giving the user different options for what they want to see and what is available to experience. This navigation cannot be so obtrusive that it seriously interferes with what they are experiencing.

Through these combined technologies we can create what computer scientists call "augmented reality," a cousin of virtual reality, in which 3D displays are used to overlay a synthesized world on top of the real world. In a situated documentary, we combine augmented reality with mobile computing and communications to embed multimedia presentations into the real world, synchronized or stabilized either to the wearer of the MJW or to objects in the real world. This means that a three-dimensional model of the real world must be created for the computer in order to synchronize the computer information with objects in the real world.

To date, we have such a detailed three-dimensional model for the Columbia University Morningside Heights campus and parts of the neighboring community. Others, including university, corporate, and governmental agencies, are developing similar 3D models for the rest of Manhattan, New York City, and many other metropolitan and other areas across the U.S. and internationally. It is a laborious process that requires a lot of computing power.

Early applications of mobile augmented reality systems (MARS) have been primarily in the areas of manufacturing, where a welder might see a schematic overlaid precisely onto the surface where a weld is needed, or in medicine, where a surgeon might obtain a 3D visualization beneath a patient's skin where an incision may be needed.[3] These types of applications normally do not require the detailed and extensive geographic information that is needed for mobile augmented reality to be used in a documentary or newsgathering situation.

A situated documentary is a form of what might best be called "context-aware" media content. The core idea is to offer on-demand text or multimedia presentations sensitive to, aware of, or tailored to the geographic, or even temporal, context. In the case of the situated documentary, audience members wear the MJW at the site of past news or historical events. Through the mobile augmented reality interface, users are immersed in a three-dimensional aural and visual re-creation of the past. In effect, it is a virtual time machine enabling the user to visit past times and events.

The situated documentary thus uses a first-person, or point-of-view, narrative structure in which the audience almost relives the events of the past, told according to basic journalistic principles such as impartiality, fairness, balance, accuracy, and using attributed sources. This is in contrast to traditional documentaries that invite the viewer to watch or observe the past as reported through the conventions of journalism in a third-person presentation on a flat or two-dimensional screen or through viewed re-enactments.

3 Situated Documentaries Produced to Date

To date, my students, working with those of Prof. Steven Feiner of the Computer Graphics Lab, have produced a series of situated documentaries demonstrating the viability of the MARS technology for the design and delivery of media content for mobile wireless communications. The first situated documentary recreated the 1968 student revolt or strike at Columbia University. Three-dimensional color-coded flags appear in world-stabilized locations indicating access points to either the 1968 revolt story or later news reports. These flags are activated simply by gazing at them for a half second or so.

Historical images, audio, and video are synchronized or stabilized against the present-day locations where they originally occurred, and users can see changes in buildings or locations by accessing virtual timelines through the MJW. This is particularly illustrated by a second situated documentary that explores the Bloomingdale Asylum for the Insane, which in the 1820s and 1830s occupied much of the space currently occupied by Columbia's Morningside Heights campus. Wearing the MJW, a user can select from a virtual timeline and see a translucent three-dimensional image of the original asylum superimposed on Low Library, transmitted in real-time via the campus wireless local area network, and move forward in time as well as space to see the 3D objects change. He or she might even enter the virtual structure, and should relevant information be available in connection to its interior, experience a further narrative from within.

My students are currently working on a situated documentary that hopes to provide a new look on the story of Edwin Howard Armstrong, a Columbia University engineering professor and the inventor of FM radio in the 1930s. The situated documentary being developed may give a new level of access to the story in the context of the place where much of Armstrong's research occurred.[4]

4 Tomorrow's Possibilities

To date, all our situated documentaries have been based on the Columbia campus and have been historical in nature. These limitations are increasingly less important, as wireless infrastructure expands and develops, especially broadband radio spectrum technologies such as WiFi (IEEE 802.11) for providing low-cost (or even free) wireless Internet access, as in the case of NYC wireless[5] and ubiquitous wireless local area networks which can provide high-speed Internet access. In September 2004, Philadelphia announced that the city was examining a project to create a city-wide wireless network.

Moreover, as mobile communications and Internet access devices proliferate and consumers grow increasingly comfortable with accessing Internet content and communications from mobile devices, there will no doubt be further opportunities to deliver to consumers context-aware media content, including news and entertainment.

There are at least three forms that context-aware news and entertainment content can take in the mobile wireless arena. First, in the near-term, context-aware media content is likely to feature breaking news or entertainment that is of a highly localized nature (i.e., information about current or recent news or cultural events specific to the location the consumer finds him or herself in) or is sensitive to the consumer's planned route, destination, preferences, demographics, or the time of day.

Marketing information is also likely to exploit the context-aware capabilities of mobile wireless communications devices. For example, whether it takes the form of spam or is delivered on demand or in exchange for time-sensitive discounts or other rewards, retail outlets might provide textual or multimedia presentations about their shops—virtual store-fronts or signs—embedded into the real-world but displayed only on wireless mobile devices. Prof. Feiner's above mentioned lab has created a virtual display for Tom's Restaurant on Broadway, the famous upper West Side diner featured in the hit television comedy series, *Seinfeld.* Walk past Tom's today wearing the Lab's mobile augmented reality system, and you'll see a virtual display overlaid near the top of the diner revealing the *Seinfeld* connection. At some point in the not-too-distant future, virtually any location in the world might boast similar virtual annotations for the appropriately equipped visitor or tourist. This information might take the form of "news you can use," where buildings and other locations or points of interest are clearly labeled and layers of additional information are available on demand. A mundane example that would likely be very popular at certain times would be information on nearby businesses that have public restrooms.

A variety of other possibilities exist in the middle-term. Imagine tapping into various meta-data that today exist in increasingly digital and online form.

Via a head-worn or hand-held display with wireless connectivity, media or other organizations might provide different types of data maps presenting three-dimensional views of those data overlaid onto the real world to which those data connect. For example, consider mapping current or historical crime data onto the locations where those crimes occurred.[6]

News organizations today routinely gather crime data on their communities, but they rarely present the data systematically or make it available to mobile wireless devices. Within a few years it may be routine for media organizations to make crime data available to mobile wireless devices via the eXtensible Markup Language (XML).

Facilitating things even further may be NewsML, Reuters publicly available variation of XML for media organizations. NewsML is a form of XML created and given away by Reuters and approved as an open standard by the International Press and Telecommunications Council.[7] In fact, Reuters is one of the leading news organizations already contemplating the design and delivery of context-aware news and information, particularly financial data. Reuters has launched a project in London to develop a next-generation editing system to produce NewsML- and XHTML-enhanced content and to make news even more targeted, customizable and platform-independent.

There is no reason why a future special archive section might not feature multimedia content and be delivered to wireless mobile devices; in fact, anyone with a PocketPC and wireless Internet access could do so today. The content would not be user or real-world synchronized or stabilized, but adding geo-referencing capabilities is not a significant technical difficulty.

5 Making the Invisible Visible

Other forms or types of information might also be displayed or overlaid upon real-world views. Among the most interesting possibilities would be to display information about the immediate or nearby environment which otherwise might be invisible to the individual. For example, imagine donning a wireless-capable head-worn display, looking skyward and seeing the size of the hole in the ozone layer, or the ultraviolet ray index, or the pollution content in the atmosphere.

In the future it may also be possible to capture video and audio in three-dimensional format for use in situated documentaries. Today, audio and video used in situated documentaries are in two-dimensional format because the cameras and microphones used typically capture only two-dimensional media. Increasingly, devices for capturing sound, pictures and video in three dimensions are being developed. As these devices develop, situated documentaries utilizing three-dimensional audio and video will make the user experience even more lifelike.

6 Unintended Consequences

Although such 3D media can make the situated documentary even more engaging and contextualized, it has the potential negative effect of blurring the line between the real and the synthetic. At present, the overlaid media are always readily identifiable by the audience member. In a future where 3D media are seamlessly produced, such differentiation may be increasingly difficult. Moreover, the long-term health or psychological impact of the use of mobile augmented reality in the real-world has not been fully tested and questions remain.

7 Conclusion

Converging wireless and wearable technologies present unique opportunities for news and information providers to create content designed for location- and other context-aware mobile devices increasingly deployed in the marketplace. Near-term opportunities include delivering breaking news and other information contextualized and customized to the location of individual audience members. Middle- and far-term opportunities include designing news and narratives that embed three-dimensional multimedia reports to mobile audiences.

Whether and how this future is realized depends on a number of factors that merit further research. Among these are: (a) whether audiences will find sufficient value in context-aware news and information to make its development commercially viable; (b) whether (and how quickly) the traditional culture of editorial production in news organizations will be adaptive enough to adopt new narrative models for news (i.e., they were generally very slow to embrace the World Wide Web as a medium for creating interactive news and are only beginning to do so now in a significant way); and (c) will there be sufficient wireless bandwidth to support the delivery of customized three-dimensional news and information on demand.

The latest high-tech gadgetry is useless if the public does not see a use for it or value in it. There are several ways in which the public may not see value in context-specific news, or see only limited or occasional value in it. If advertising becomes too obtrusive people may avoid using mobile wireless news services. Likewise, content providers must constantly be aware of what types of content best suits context-specific formats and current limitations on display technologies. It is likely that adoption may occur in fits and starts, perhaps being used in very limited contexts initially while other contexts fail miserably. As the public becomes more comfortable with using wireless and wearable technologies, they will likely see more ways to utilize them.

News organizations have been particularly slow in adopting and fully utilizing digital tools and the Internet in their newsgathering and news production process. The changes in news consumption that can be achieved with wireless and wearable computing promises to be as radical, if not more so, than the changes the computer and Internet have brought to the relationship between audience and journalist.

Endnotes

1 Spratling, R. (December 3, 2002). Interview conducted via email.

2 The MJW includes a see-through head-worn display, an orientation tracker which uses a gyroscope to monitor the user's orientation in the environment (allowing the user to select interactive objects displayed in view simply by looking at them, or what is called gaze approximation), a wearable computer capable of processing three-dimensional graphics as well as a hand-held PC, a differential global positioning system (GPS) receiver accurate to within about 1.5 centimeters, and a spread spectrum radio communication link providing high-speed (11 megabits per second) wireless Internet access (http://www.cs.columbia.edu/graphics/projects/mars/mars.html). Prof. Feiner describes the technology making up his laboratory's mobile augmented reality system in an April 2002 Scientific American article (http://www.sciam.com/2002/0402issue/0402feiner.html).

3 More information on medical applications of augmented reality is available at www.cs.unc.edu/~us/; other AR research applications can be found at www. augmented-reality.org. Mobile augmented reality systems are emerging for a variety of applications, ranging from the military battlefield (http://www.popsci.com/popsci/computers/article/0,12543,190327-1,00.html) to the consumer marketplace. In addition to ours, some of the earliest trials have involved multimedia museum tours in which geographic range is limited.

4 In addition to the situated documentaries mentioned above, students in previous years also created two other situated documentaries. One was on the connecting tunnels under the Columbia University campus (used both by asylum inmates and rebellious students to escape inclement weather but now only accessible to maintenance personnel). Another was on the groundbreaking work done by physicist Enrico Fermi, who created the first atom smasher underneath the Columbia campus.

5 See: http://www.nycwireless.net/.

[6] On February 14, 1972, investigative reporter David Burnham reported in *The New York Times* the first crime map of New York City broken down by police precinct. Burnham's innovative report provided data for the first time on crime by precinct as well as by type of crime, and was made possible by the then-unprecedented use of the computer as a tool to analyze public records.

[7] See: www.iptc.org.

References

Anonymous. (December 30, 1916). Wireless transmission of news. *Telephony,* (pp. 32-33). Retrieved from the World Wide Web: http://www.angelfire.com/nc/whiteho/1916df.htm.

Associated Press. (1940). The Story of News. (Copyright 1940.)

Azuma, R.T. (1997). A survey of augmented reality. *Presence: Teleoperators and Virtual Environments, 6*(4), 355-385. Retrieved from the World Wide Web: www.cs.unc.edu/~azuma/ARpresence.pdf.

Azuma, R. A., Baillot, Y., Behringer, R., Feiner, S. K., Julier, S., and MacIntyre, B. (2001). Recent advances in augmented reality. *IEEE Computer Graphics and Applications, 21*(6), 34-47. Retrieved from the World Wide Web: www.cs.unc.edu/~azuma/cga2001.pdf.

Ditlea, S. (2002). Augmented Reality. *Popular Science Online.* Retrieved from the World Wide Web: http://www.popsci.com/popsci/computers/article/0,12543,190327-1,00.html.

Feiner, S. (2002). Augmented reality: a new way of seeing. *Scientific American.* Retrieved from the World Wide Web: http://www.sciam.com/2002/0402issue/0402feiner.html.

The New York Times (2002). The Laramie Project Archives. Retrieved from the World Wide Web: http://www.nytimes.com/ads/marketing/laramie/index.html.

Pavlik, J. & McIntosh, S. (2003). *Converging media: introduction to mass communication in the digital age.* Boston, MA: Allyn & Bacon.

8
Mobile Peer-to-Peer Content and Community Models

Valerie Feldmann

1 Introduction

Mobile peer-to-peer applications are being envisioned as exciting new content applications for next generation's mobile wireless networks. Peer-to-peer applications on the fixed-line Internet are experiencing strong demand and rapid diffusion. However, wireless peer-to-peer applications are likely to be very different.

Mobile peer-to-peer applications are still a predominantly conceptual phenomenon. Successful applications are not easy to imagine. Mobile phones have limited storage and restricted input capabilities, making them inconvenient to initiate queries; prices for mobile data transmission are too high for customers; and while Internet file sharing offers the huge databases of other users around the globe, mobile file sharing may rather be restricted to local environments, for example, to ad hoc social networks.

These factors suggest that mobile peer-to-peer file sharing will be different from traditional fixed-line file sharing and will serve different needs. This chapter will discuss what value mobile communications can add to peer-to-peer applications and vice versa. The analysis will focus on the technological characteristics of mobile peer-to-peer applications, content models between user generated and professionally produced content, the concepts of identity and community in the context of mobile communications, and media companies' challenges and options to effectively manage mobile peer-to-peer communities. A discussion of emerging policy issues in the context of mobile peer-to-peer communities concludes the chapter.

2 Technological Characteristics of Mobile Peer-to-Peer Applications

Mobile peer-to-peer applications via handheld devices differ technologically from fixed-line desktop PC or notebook peer-to-peer applications. Dennis and Ash (2001) suggest that technological platforms are the dominant feature of the identity of new media. Therefore, this section will analyze the technological characteristics.

2.1 Definition and Categorization of Mobile Peer-to-Peer Applications

Peer-to-peer applications in fixed-line environments are defined as a range of applications that harness the free resources available at the edges of the Internet such as storage, cycles, content, and human presence (Shirky, 2001).[1] Three applications helped define the strengths of fixed-line peer-to-peer: Napster, which created a peer-to-peer file sharing system; SETI@home, an initiative of the Search for Extraterrestrial Intelligence which turns a group of disparate computers into a supercomputer; and ICQ, the original instant messaging program (Shirky, 2002).

Mobile peer-to-peer applications can be defined as a range of applications that harness (a) ad hoc interaction through human presence and physical proximity; (b) virtual storage through distributed content availability on mobile Internet appliances in both public and private environments; and (c) shared broadband access through bandwidth sharing. The devices used for peer-to-peer applications are expanding from stationary to mobile. The PC and other stationary devices have large processing and storage capabilities. Mobile devices, however, have limited processing, battery, and storage capabilities. They are, for example, not suited for building a personal media library. Similarly, display size and input facilities of a mobile information device do not offer convenient browsing and searching opportunities in mobile settings. Moreover, cellular networks do not offer the bandwidth needed to transfer large files and is expensive (see Brown, in this volume). For a systematic discussion of the challenges and potentials of mobile peer-to-per applications it is useful to categorize potential applications.

Fattah (2002) offers a categorization of peer-to-peer applications that differentiates active applications and idle utilizations. Active applications comprise user collaboration such as file sharing, gaming, and Electronic Data Interchange (EDI). Idle utilizations consist of resource utilizations such as bandwidth conversion and of supercomputing for high-performance applications. Wireless peer-to-peer applications can follow this differentiation into active and idle utilizations (see Figure 1). Active applications comprise user collaboration such as mobile ad hoc file sharing, mobile instant messaging, or mobile multiplayer gaming as well as application interaction such as Bluetooth-enabled data exchanges. Idle utilizations comprise resource utilizations

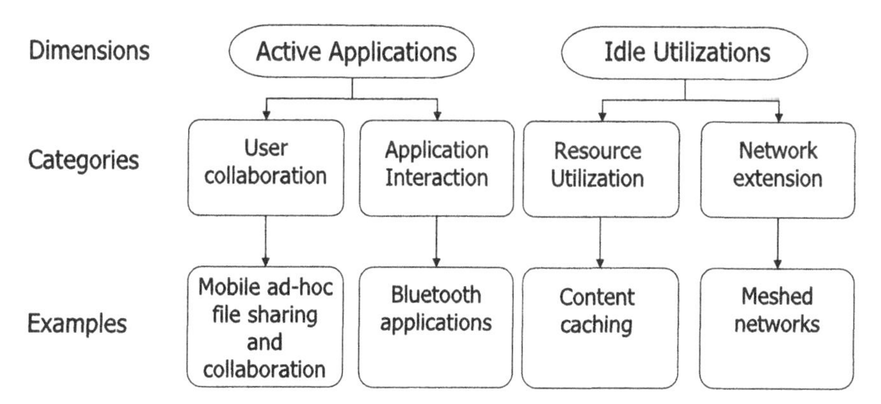

Figure 1: Categorization of wireless peer-to-peer applications
Source: adapted from Fattah (2002), p. 23

such as caching content and the extension of wireless networks for meshed networks.

Idle utilization applications may become more promising for wireless peer-to-peer applications. Yet, for the discussion of the role of mass media content for mobile peer-to-peer applications we will focus, for the remainder of this chapter, on the first application category of user collaborations.

2.2 Mobile ad hoc Networks

Mobile sharing applications may become available through wireless ad hoc networks based on physical proximity. Mobile ad hoc networks are decentralized and self-organizing. They are dynamic and continuously reshaped into multiple clusters. These mobile devices interact as autonomous peers (Kortuem et al., 2001; Gruber, Schollmeier, & Kellerer, 2004). Personal area networks (PANs) are a special category. They are low power and low range wireless networks that connect personal mobile devices or wearable computers. The devices function as nodes for peer-to-peer networks or as hosts for the most frequently accessed information.

The focus of data sharing can, for example, lie on increasing the data availability to users roaming a metropolitan area (Papadopouli/Schulzrinne, 2001; Wolfson, Xu, & Sistla, 2004). In this case, the networks need no support of any infrastructure for data dissemination among the mobile devices. Thereby, they overcome intermittent connectivity to the Internet and allow location-dependent and collaborative services. The relevant parameters are the density of cooperative hosts and their mobility. Particularly in urban environments users increasingly demand ubiquitous data availability. In these environments, mobile ad hoc peer-to-peer networks can be used to cache popular content.

Mobile ad hoc information systems are particularly interesting, because personal area networks allows integrating mobile devices into everyday social interaction. Peer-to-peer applications for proximity-aware mobile collaboration can augment social encounters and face-to-face interactions. However, in mobile ad hoc networks, time becomes a critical resource. The exchange of information is bound to happen fast as physical presence is required. If a transfer takes too long, the network connection might be interrupted. Therefore, Kortuem et al. (2001) suggest to exchange URLs that point to files on a server rather than the files themselves.

3 Mobile Peer-to-Peer Content Models

The basic premise of sharing applications is that consumers have something valuable to share. It is useful for the discussion of content models to differentiate two organizational levels of content production: professionally produced, user-generated content. Each will be discussed in turn.

3.1 User-Generated Content

Personal media files such as digital pictures, voice recordings, or short video sequences may offer the highest incentive to swap digital files with peers and friends. Since storage is not a distinctive feature of mobile personal devices, user-generated content may be transient. However, mobile peer-to-peer platforms can serve as virtual storage that makes personal media files ubiquitously and instantaneously accessible.

Sharing of personal data as an expression of connectivity with peers and friends may become an essential revenue driver for mobile operators. Communications and message services already produce substantial revenue. Connectivity has always mattered more than (professionally produced) content (Odlyzko, 2001). In the 19th-century, for example, postal services derived their profits from letters and subsidized newspaper distribution. E-mail is creating a lot of value in the Internet, although its popularity was not foreseen by the ARPANET's planners. The perceived prominence of connectivity over content can be used today to explain the widespread adoption of SMS and raise questions about premium data services of next generation 3G networks. Odlyzko (2005) even suggests that the main role of 3G wireless systems should be to stimulate voice usage. However, the willingness to pay for connectivity may also extend to sharing personal media files.

An essential incentive for personal file sharing may lie in the user's identity representation within social networks. A distinguishing element of mobile environments as opposed to virtual environments on the fixed-line Internet is that the identity of the user is rather enforced than blurred. The sociologist Sherry Turkle (1997) has suggested that users on the fixed-line Internet deliberately choose multiple identities that include gender switch and different communication intentions. In contrast, the mobile phone with its distinct telephone number and SIM card that stores personal information reinforces the identity of its user. The personality representation in mobile wireless environments is given via the device that can represent user and lifestyle (Pedersen, Nysveen, & Thorbjørnsen, 2003). Personalized style elements include the cell phone, but also content elements such as certain ring-tones or icons as personality representations when sending an SMS. Thus, the concept of identity is of particular interest in mobile communications environments (Feldmann, 2005).

Research on SMS suggests that mobile communications is predominantly used to maintain existing personal relationships with a rather small group of peers. Youth send SMS in more than 50% of all cases to partners or best friends; they rarely write to family members or strangers, both below 10% (Hoeflich & Roessler, 2001). Another study reveals that youth exchange SMS regularly with 1-3 persons in 40% of the cases, with 7 to 9 persons in 20% and with more than 10 persons in only 4% of the cases (Schlobinski et al., 2001). A study surveying 9 European countries (Smoreda & Thomas, 2001), shows that contact patterns are strongly concentrated geographically, whereas Internet-based written contact is far more dispersed.

This user behavior suggests that file sharing via mobile phones may be particularly interesting for user-generated media. Such file sharing is different from the fixed-line global file sharing among a group of anonymous users. Identity construction in personal networks may become a strong proposition for mobile peer-to-peer usage. Such a development would also avoid many of the intellectual property right issues of stationary peer-to-peer file sharing.

3.2 Professionally Produced Media Content

In the fixed line Internet, music, movies, and TV series are the most popular media content for file sharing. For mobile devices, however, only a small amount of mobile media content is available at the moment that could be used to create a mobile peer-to-peer community. Mobile content and services are subject to licensing agreements between media companies and mobile operators and usually not intended to allow content sharing. The costs of securing these mobile content rights, for example, for sports highlights, are not negligible and they are currently limiting the media content and services carriers can offer.

In the light of this background, three options for mass media content may emerge: (1) promotional sample contents; (2) user-contextualized content, and (3) branded content.

Sample content is a viable sharing option for professionally produced content that can be used as a promotional tool. It may become linked with subsequent purchasing options. For example, music firms could release twenty seconds of a new song that is available for sharing and of delivering the entire song for digital download after receipt of a certain fee sent by a minimum number of users (Dolan, 2000). A promotional campaign by the publishing house Simon and Schuster in New York City provides an example. Telephone kiosks on the streets of Manhattan offered to beam excerpts from the latest short stories by Stephen King to consumers' personal digital assistants. Loyal readers or other interested audiences could download approximately 400 words and share the sample with other owners of handheld computers (Elliott, 2002).[2] Music magazines could provide editorial content on artists when songs are purchased and downloaded on a mobile personal device.

The integration of personal messages and professionally produced content may provide another sharing option when mobile users receive the opportunity to make pieces of purchased mobile media available to friends and to comment on it. Such user-contextualized content may become subject to sharing among peers as well. Other forms of user contextualization may develop such as self selected and bundled songs that users share as a link list.

A third option is to make branded content available for sharing. Branded content is produced and financed by advertisers. It is attractive for users and it goes beyond advertising by disseminating real content (ECC, 2000). Firms from the consumer goods industry already produce content professionally as a means for brand building or customer relationship management. They could, for example, produce mobile games for sharing. If different games are available, consumers may appreciate access to the virtual storage capacities of mobile peer-to-peer communities due to device storage limitations.

In all cases of promotional sample content, user contextualized content, and branded content it is in the interest of the (media) company when users share these files. Full digital media files such as MP3 files may become interesting for mobile peer-to-peer platforms when the cost of transmission can be reduced and when peer-to-peer business models will be developed. Kortuem et al. (2001) suggest three forms of impromptu collaboration for mobile music MP3 file sharing: (1) face-to-face file sharing during personal encounters; (2) personal agents that act on behalf of users; (3) and institutionalized file sharing platforms that include authentication, security, and payment transactions. A commercial mobile peer-to-peer sharing model is likely to involve a digital rights management (DRM) system for mobile content. In this scenario, media companies could develop an institutionalized peer-to-peer platform that may allow paid content models. Figure 2 summarizes these two models.

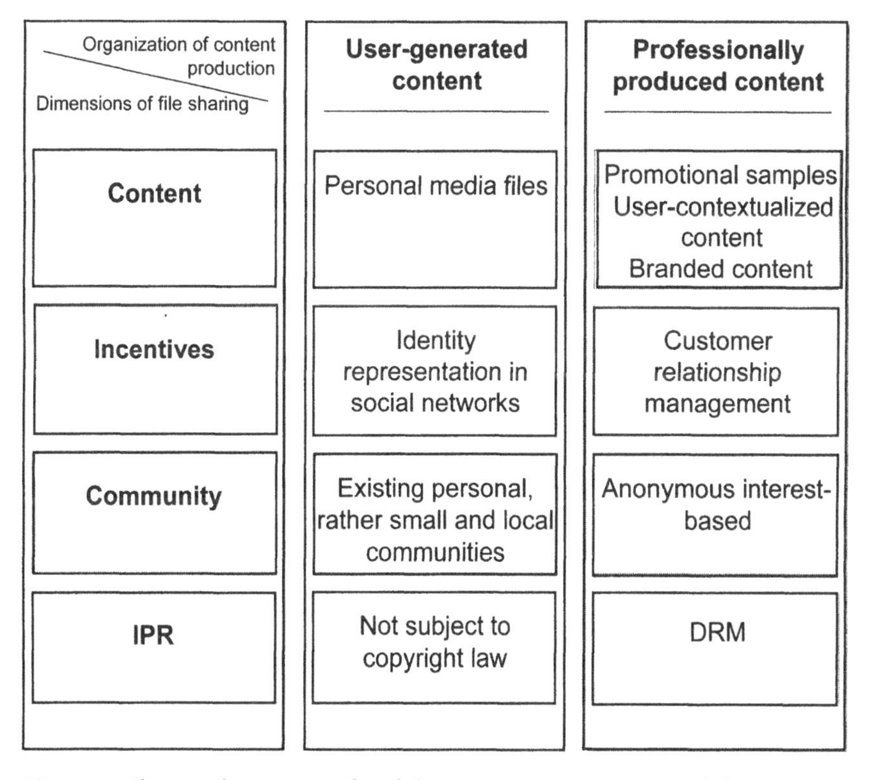

Figure 2: Sharing dimension of mobile peer-to-peer content models

4 Mobile Peer-to-Peer Community Models

Community concepts, distinguished by the common values and common rules, shared resources and shared purposes that provide reason for the community, are a core construct in the social sciences. Mobile communities can be subject to ad hoc community formation (Rheingold, 2002); they can also be envisioned as extensions of real-life geographically based communities (Hollander, 2000) and of geographically dispersed virtual communities (Rheingold, 1995). For peer-to-peer applications this distinction may make a difference for community management strategies since peer-to-peer communities are often characterized by low community commitment and free riding problems.

4.1 Mobile Community Evolution

Mobile communities can be distinguished from online communities according to social scope and formation. Mobile communities can be both communities of anonymous members or a group of peers from an existing social network. When they relate to existing social networks, identity building within the social network may be an incentive for sharing and contributing to the community of friends since an essential reason for mobile phone and mobile community usage is to stay in close contact to friends and peers. Thus, buddy systems as implemented in instant messaging systems may become an interesting option for mobile community building. Mobile communities among anonymous members may form around shared interests such as lifestyle or health issues.

In the case of mobile ad hoc community formation, mobile communities are location dependent. One essential implication is that they enable consumers to act upon impulses. In the context and tradition of consumer research, impulse as opposed to habituated consumption decisions are subsumed under consumer decisions with low cognitive control which often involve emotions as well as an immediate action response to a stimulus (Stern, 1962; Rook, 1987; Kroeber-Riel & Weinberg, 1999).

Media companies have a longstanding tradition in building communities around their content. The first imagined community (Anderson, 1991) may have emerged with the subscriber base of a newspaper. Communities have also been introduced as a model for media content management on the Internet. The participants in these content communities create the context of the professionally produced content (Hummel/Lechner, 2001).

Content communities often evolve around strong media brands and can therefore be classified as brand communities. A brand community as introduced by Muniz & O'Guinn (2001) is a specialized non-geographically bound community that is based on a structured set of social relationships among admirers of a certain brand. Brand communities embrace the qualities from the concept of communities such as shared consciousness, rituals and traditions. For strong media brands, computer-mediated environments support the community creation around mass media content; individual users can devote messages, links, or Web pages to the community. Thereby, brand communities socially construct and shape media brand identities. Media companies may now expand their community models to mobile ad hoc communities around content versioning and ubiquitous media brand contact opportunities.

4.2 Mobile Peer-to-Peer Communities

Communities do not have to be formed around a belief or value, they can also be formed around an activity. Peer-to-peer content communities on the Internet have often evolved around music and movie files that can be searched and downloaded. However, it is debatable if a common use of an application can be called a community. The sociologist Brint (2001) offers the differentiation of interest based and activity based communities that is useful in this context. The formation around peer-to-peer file-sharing platforms is more activity based than belief based. Members share resources and purposes rather than common values. The personality representation, for example, is based on a nickname, the connection type, the number of shared files, and the (music) file properties. This user profile creates a sort of cultural capital that can be further segmented into sub-cultural capital when the specificity of the content raises (Poblocki, 2001). For example, in the Napster community, users with sub-cultural capital often gather in so called Hot Lists. However, the bonds between members are rather weak and fugitive. There is little incentive to develop stronger ties since individual goals can be attained without them. This results in a low sense of commitment and a weak brand loyalty towards the peer-to-peer community. Loyalty to Napster has decreased immensely after files were banned and the selection decreased (King, 2001).[3] Thus, peer-to-peer communities can be classified as activity based communities that exert weak ties on its members. The same may hold true for mobile peer-to-peer communities that represent (anonymous) ad hoc community members; in the case of a group of friends that constitute a mobile peer-to-peer community the community loyalty is suggested to be higher due to strong ties from existing social networks (see Figure 3).

If file sharing is based on physical proximity in mobile peer-to-peer communities, the context of the file sharing is essential. Whereas PC-based peer-to-peer content sharing is executed with anonymous partners within the private space, trading partners are aware of each other when people come face-to-face since PAN exchanges only happen across short distances within close physical proximity. Physical proximity includes observing certain social clues, possibly talking, and it has consequences on politeness and trust (Kortuem et al., 2001). Since file sharing is based on short personal encounters or a possibly unstable mobile wireless network connection, the files should be of small size in order to be able to complete any sharing activity. In addition to volatile network connections, user attention in nomadic Internet environments is an even scarcer resource than in the stationary Internet; it might be occupied by other real-world tasks and unintentionally interrupt sharing processes.

Community Model / Criteria	Fixed-line peer-to-peer community	Mobile ad hoc peer-to-peer community
Community elements	Activity-based location independent	Belief- / activity-based location (in)dependent
Personality representation	Nickname, connection type, number of shared files, file properties	Personal and personalized device represent users and lifestyle
Bonds between members	Weak, fugitive ties	Personal ties / weak, fugitive ties
Brand loyalty	Low sense of commitment	Strong sense of commitment in the case of personal ties
Sharing context	Stable	Volatile Impulsive

Figure 3: Characteristics of fixed-line and mobile peer-to-peer communities

The prominence of interpersonal communication management via the mobile phone suggests that an integration of professionally produced and user-generated content may be beneficial for the deployment of mobile communities. Mass media companies that intend to build mobile peer-to-peer communities may be able to unlock revenue opportunities by integrating pieces of their copyright protected content by means of interpersonal communications and user-generated content. For example, they can let users bundle music albums of self-selected (and potentially self-produced) songs they can share. When users purchase songs for download on their mobile device sound greeting to friends would be allowed.

Some prototype content models have been tested for mobile file sharing that integrate elements from interpersonal communications. In 2002, Orange Sverige, BMG Sweden, Compaq, and the IBM e-business Innovation Center developed a 3G music prototype (van Impe, 2002). It functioned like a music encyclopedia and personal jukebox.[4] The pilot test 'x-files' allowed 30 test users to use the service for one day. A client server architecture that integrated a digital rights management system allowed consumers to buy digital licenses. Users were able to select, listen to, search, organize and exchange music, video, voice, and text content. Parts of the songs could be sent to friends via IM or e-mail. Recording functions for user-generated sounds could be activated by dialing a certain number and record the sound. Whereas older test users tended to be more interested in a jukebox application, younger test users preferred to play with the application and communicate it immediately to their friends. Orange sees its own core value in creating communities around that content (Neumann, 2002). Another emerging mobile peer-to-peer application is the Japanese Gnutella project, http://jnutella.org. It is deploying Gnutella on i-mode mobile phones where the results of a search are tailored to mobile phone interfaces.

File sharing of mobile mass media content that integrates forms of interpersonal communications between members of a social network may become a promising model for mobile communities. Yet, mobile file sharing may have its greatest potential within a small world social network based mobile community in order to exchange personal media files.

4.3 Community Management of Mobile Peer-to-Peer Communities

In peer-to-peer communities, free riding is a fundamental problem. Almost 70% of Gnutella users share no files, and the top 1% of sharing hosts return nearly 50% of all responses (Adar & Huberman, 2000). This phenomenon is related to the problem of securing enough cooperation in large and anonymous systems.[5]

A possible solution to free riding is to restrict membership for mobile peer-to-peer content communities, for example, limit it to mobile buddy lists. Asvanund et al. (2001) analyze the optimal size of a mobile wireless peer-to-peer network and suggest multiple small networks instead of a single monolithic network. In their model, they take three elements into account: the amount and desirability of content provided by the user, the size and frequency of downloads initiated by the user, and the size of networks in terms of users and capacity. They model positive network externalities as new selections of content provided by additional users. Negative network externalities are modeled as network congestion caused by additional users.[6] Marginal value of additional users is declining and the marginal cost is increasing with the number of users of a monolithic network.

Another option for mobile peer-to-peer community management is to set up a market based architecture that allows peers to buy and sell resources such as processing resources or bandwidth capacity. This is possible through the introduction of micro-payments (Golle, Leyton-Brown, & Mironov, 2001). The imposition of financial transfers abstract from altruistic reasons for sharing and contribute to reaching network equilibrium. Micropayment mechanisms can, for example, reward users for uploading and charge them for downloading.

Pricing has also been applied to transit traffic in wireless peer-to-peer networks (Chandan & Hogendorn, 2001). In these peer-to-peer network models each user's mobile device must dedicate some of its bandwidth and battery power to facilitating the transit traffic of other users. The findings suggest that organizing peering through a club may be the best solution to possible congestion problems. It could internalize the network externalities by instituting an entry fee or a limitation on bandwidth use. The club could be run for profit or it could be a voluntary association with a restricted number of slots.

The implications for wireless operators and media content providers that derive from these characteristics of mobile peer-to-peer community management suggest to abandon generalist content strategies and to specialize content offers. Mobile peer-to-peer content communities are likely to introduce optimal membership rules as well as pricing mechanisms to avoid free riding and congestion externalities.

5 Emerging Policy Implications

Two dominant policy issues emerge, the first affecting copyright issues, the other one addressing security and privacy.

Mobile DRM systems are currently developed by mobile operators, content providers, and independent third parties. The discussion about copyright infringement from fixed-line peer-to-peer applications (Greenstein, 2001; Picot, 2004) extends on wireless and mobile networks and devices when professionally produced media content will be available for download to mobile devices as well as for mobile streaming. Yet, when mobile content is regarded as a means to support and expand revenue generation in the core media brand offer, media companies may have an incentive to allow for open platforms and mobile sharing processes. Moreover, consumer choice on the use of digital media files on more than one device is an essential concern from a fair use perspective. Another issue is the liability of third parties, since existing copyright case law imposes liability on third parties for the infringing act of others (Greenstein, 2001). If a mobile user happens to become a frequently used node in an ad hoc network this user may be regarded as a (temporary) central authority.

Mobile peer-to-peer community models also raise data security and privacy concerns with opening a mobile device to the public for accessing or routing data. There is increased danger of mobile virus spread and data theft with open mobile devices. These concerns may contribute to favor restricted mobile peer-to-peer community access. When more sensitive data, for example, transaction and payment information, is stored on mobile devices security concerns add to the privacy issues. In the case of mobile payments for file sharing, contracts also need to consider the dynamically changing network topography and potential transfer failures.

6 Conclusion and Outlook

Mobile peer-to-peer communities face technological challenges such as restricted resources on the mobile computing device as well as narrow bandwidth and unreliability of the wireless link. Yet, instant and ubiquitous access to virtual storage or ad hoc community formation may offer incentives for file and information sharing. A mobile Napster is unlikely to emerge due to the constraints of mobile devices and network capacities. Yet, the social dimensions of mobile peer-to-peer communities are interesting for the discussion of alternative applications. They may take advantage of the ad hoc community formation capabilities, social networks of mobile users, social aspects of physical proximity, and new interaction patterns.

Different content models exert much influence on user's incentive to share and contribute to mobile peer-to-peer communities as well as their behavioral patterns. User generated content such as digital pictures, sound greetings, or video recordings may become popular properties for sharing within social networks. Since the mobile phone is increasingly used as a means of self-expression, its nature suggests personal file sharing and using mobile peer-to-peer platforms as a means of virtual storage. Mobile operators have an interest in these forms of exchange, because it drives mobile data traffic to their networks. Professionally produced media content, on the other hand, may be shared in the form of links that point toward files that will be available for later download on a different device or via a different network. Media companies can organize sample content sharing via mobile peer-to-peer communities in order to strengthen brand loyalty and revenues in other offline and online media channels. Branded content, media content that is financed by an advertising client, may offer an even better proposition for mobile peer-to-peer sharing since the advertiser is interested in the viral effects from mobile users.

Free riding problems in peer-to-peer environments suggest the need for community management. Rules for sharing could consist of micro-payment models or membership restrictions. According to user preferences, mobile peer-to-peer communities can be efficiently managed in the form of clubs based on special interests. However, in mobile peer-to-peer communities that are activity based and characterized by weak ties, free riding may still be a problem.

The role of mass media providers can lie in the community platform provision and management of mobile peer-to-peer communities. It is limited for user generated content models since mobile operators will have a far greater incentive to provide community platforms for these content models. In the case of professionally produced content, however, media companies can play a role in using their brand strength to offer community platforms to targeted user groups. Their strongest proposition is the cross-media integration of mobile community models with their existing online communities and the stimulation of cross-media audience flows between different media. When media companies manage to build a marketplace for peer-to-peer community models, emerging cross-network approaches (Feldmann, 2005) are yet another alternative that will offer users more choice with regard to price and quality of service.

Endnotes

[1] Because these resources are subject to changing IP addresses, peer-to-peer transcends the Domain Name System (DNS) layer and reverts control back to the PCs. That distinguishes this definition from the peer-to-peer definition in the 1970s and 1980s. Then, peer-to-peer technology connected mainframe computers that were permanently connected and had a permanent IP address. This changed with the growing number of PCs connecting to the Net, see Fattah (2002), p. 20.

[2] This sharing process did not happen via a mobile wireless peer-to-peer network yet, but with the conventional PDA capability of beaming content.

[3] In April 2001, Napster use fell by nearly 36 percent from the previous month, according to a study by Webnoize research, see King (2001).

[4] The following information is based on a telephone interview with Orange's business development manager Frederick Neumann, conducted on March 27, 2002.

[5] Napster users have even been witnessed to misrepresent the speed of their network connection in order to discourage other users from connecting to them, see Adar & Huberman (2000), p. 7.

[6] Research suggested positive network externalities for song availability with up to 8,000 users. Negative network externalities increased exponentially as the number of users approached the hypothesized capacity of the network, see Asvanund et al. (2001), p. 3.

References

Adar, E. & Huberman, B.A. (2000). Free riding on Gnutella. *First Monday, 5*(10).

Anderson, B. (1991). *Imagined communities: reflections on the origin and spread of nationalism.* London: Verso.

Asvanund, A., Clay, K., Krishnan, R., & Smith, M. (2001). Bigger may not be better: an empirical analysis of optimal membership rules in peer-to-peer networks. *Telecommunications Policy Research Conference,* Alexandria, VA, October 27-29, 2001.

Brint, S. G. (2001). Gemeinschaft Revisited: A Critique and Reconstruction of the Community Concept. *Sociological Theory, 19*(1), 1-23.

Chandan, S. & Hogendorn, C. (2001). The bucket brigade. Pricing and network externalities in peer-to-peer communication networks. T*elecommunications Policy Research Conference,* Alexandria, VA, October 27-29, 2001.

Dennis, E. E. & Ash, J. (2001). Toward a taxonomy of New Media – Management views of an evolving industry. *The International Journal on Media Management,* 3(1).

Dolan, D. P. (2000). The big bumpy shift: digital music via mobile Internet. *First Monday, 5*(12).

ECC. (2000). *E-conomics.* Springer: New York.

Elliott, S. (March 19, 2002). Stephen King's new book is on the beam, literally. *The New York Times.*

Fattah, H. M. (2002). *P2P: how peer-to-peer technology is revolutionizing the way we do business.* Chicago: Dearborn Trade Publishing.
Feldmann, V. (2005). *Leveraging mobile media. Cross-media strategy and innovation policy for mobile media communication.* Heidelberg: Physica (forthcoming).
Golle, P., Leyton-Brown, K. & Mironov, I. (2001). *Incentives for sharing in peer-to-peer networks.* Retrieved from the World Wide Web: http://crypto.stanford.edu/~mironov/papers/welcom01.pdf.
Greenstein, S. (2001). Copyright in the age of distributed applications. In B.M. Compaine and S. Greenstein (Eds.), *Communications policy in transition: the Internet and beyond* (pp. 369-396). Cambridge: MIT Press.
Gruber, I., Schollmeier, R. & Kellerer, U. (2004). Performance evaluation of the mobile peer-to-peer service. Retrieved from the World Wide Web: http://www.lkn.ei.tum.de/lkn/mitarbeiter/hrs/Komponenten/paper/MPP_camera.pdf.
Hoeflich, J. & Roessler, P. (2001). *Forschungsprojekt 'Jugendliche und SMS – Gebrauchsweisen und Motive. Zusammenfassung der ersten Ergebnisse.* Retrieved from the World Wide Web: http://www.uni-erfurt.de/kw/forschung/smsreport.doc.
Hollander, E. (2000). Online communities as community media. A theoretical and analytical framework for the study of digital community networks. *Communications* (25), 371-386.
Hummel, J. & Lechner, U. (2001). The Community Model of Content Management – A case study of the music industry. *The International Journal on Media Management, 3*(1).
Impe, M. van (January 25, 2002). *Orange sverige develops mobile Napster.* Retrieved from the World Wide Web: http://www.mobile.commerce.net/story.php?story_id=1193.
King, B. (May 2, 2001). RIAA head: Napster is done. *Wired,* Retrieved from the World Wide Web: http://www.wired.com/news/print/0,1294,43487,00.html.
Kortuem, G., Schneider, J., Preuitt, D., Thompson, T. G. C., Fickas, S., & Segall, Z. (2001). When peer-to-peer comes face-to-face: collaborative peer-to-peer computing in mobile ad hoc networks. Retrieved from the World Wide Web: http://www.cs.uoregon.edu/research/wearables/Papers/p2p2001.pdf.
Kroeber-Riel, W. & Weinberg, P. (1999). *Konsumentenverhalten.* München: Vahlen.
Muniz, A. M. Jr. & O'Guinn, T. C. (2001). Brand community. *Journal of Consumer Research, 27,* 412-432.
Neumann, F. (2002). Personal communication, March 27.
Odlyzko, A. (2005). Talk, talk, talk. So who needs streaming video on a phone? In ECC (Ed.), *E-merging media. Communication and the media economy of the future* (p. 170). Berlin: Springer.
Oram, A. (Ed.). (2001). *Peer-to-peer. Harnessing the power of disruptive technologies.* Sebastopol: O'Reilly.
Papadopouli, M. & Schulzrinne, H. (2001). *Performance of data dissemination among mobile devices.* Retrieved from the World Wide Web: http://www1.cs.columbia.edu/~library/TR-repository/reports/reports-2001/cucs-005-01.pdf.
Pedersen, P. E., Nysveen, H., & Thorbjornson, H. (2003). Identity expression in the adoption of mobile services: the case of MMS. *SNF Working Paper* (26/03). Foundation for Research in Economics and Business Administration.

Picot, A. (2004). *Digital rights management.* Berlin: Springer.
Poblocki, K. (2001). The Napster Music Community. *First Monday, 6*(11).
Rheingold, H. (1995). *The virtual community: Finding commection in a computerized world.* London: Minerva.
Rheingold, H. (2002). *Smart mobs: the next social revolution.* Cambridge: Perseus.
Rook, D.W. (1987). The buying impulse. *Journal of Consumer Research, 14,* 189-199.
Schlobinski, P., Fortmann, N., Gross, O., Hogg, F., Horstmann, F., & Theel, R. (2001). *Simsen. Eine Pilotstudie zu sprachlichen und kommunikativen Aspekten in der SMS-Kommunikation.* Retrieved from the World Wide Web: http://www.mediensprache.net/networx/networx-22.pdf.
Shirky, C. (2001). Listening to Napster. In Oram, A. (Ed.), *Peer-to-peer. Harnessing the power of disruptive technologies* (pp. 21-37), Sebastopol: O'Reilly.
Shirky, C. (2002). Foreword. In Fattah, H. M., *P2P: how peer-to-peer technology is revolutionizing the way we do business.* Chicago: Dearborn Trade Publishing.
Smoreda, Z. & Thomas, F. (2001). Use of SMS in Europe. *Usages,* 10, Newsletter of France Telecom R&D.
Stern, H. (1962). The significance of impulse buying today. *Journal of Marketing,* 59-62.
Strauss, N. (February 18, 2002). Record labels' answer to Napster still has artists feeling bypassed. *The New York Times.*
Turkle, S. (1997). *Life on the screen.* New York: Touchstone.
Wolfson, O., Xu, B., & Sistla, A. P. (2004). An economic model for resource exchange in mobile peer-to-peer networks. Retrieved from the World Wide Web: http://www.cs.uic.edu/~wolfson/p2p_ps/ssdbm04-final.pdf.

9
Contents and Services for Next Generation Wireless Networks

John Carey

1 Introduction

Next generation wireless services are very broad in scope. They include a range of user devices, for example, a new generation of cell phones, personal digital assistants (PDAs) with voice and wireless data capabilities, laptop computers equipped with cards to receive and send wireless data, and a number of hybrid devices. They also include many ways in which wireless networks can be configured, including 2, 2.5, and 3G wireless networks, as well as wireless local area networks such as WiFi. In order to manage the scope of issues associated with wireless networks, this chapter excludes wireless networks that are used exclusively within an office building or home.

What services do people want, what will they pay for, and how can innovative content be fostered in the new wireless environment? Further, what we can learn from existing behavior with wireless devices, in the U.S., Europe and Asia, to inform the design and development of new services? It is also useful to ground the analysis with an understanding of some historical patterns of new technology adoption and the process by which cellular telephone service moved into society. Will the next generation of wireless technology and services follow the same path of evolution as the cell phone: from business to consumer; from low volume to high volume usage; and from high-priced, purposeful communications and content with tangible value to lower-priced general communications and playful content? Some developers of content for 3G services believe that playful or entertaining content will be among the earliest applications that will be adopted (Dvorak, 2004).

2 Adoption and Use of Current Wireless Services

The cell phone has been widely adopted in many societies over the last decade and become a core component in everyday life for hundreds of millions of people. In 1990, the ratio of wired to cellular phones was 50:1. By 2000, the ratio had narrowed to 3:2, and by 2002 the number of cell phones exceeded wired lines world wide (ITU, 2004).

2.1 Changing Uses of Cell Phones

The ways in which cell phones are used has also changed over time. One change has been that more people have kept their cell phone on for long periods of time, to receive calls. In the U.S., this is an artifact of billing. Since people pay for both sent and received calls, many people used cell phones only to send calls when the price was high. Some in this group used pagers to receive messages, which they then returned using their cell phone. As the price dropped, more people used cell phones to both send and receive calls (Katz, 1999). This had an important impact on the so-called networking effect. That is, the value of a network is related in part to the number of people you can reach. Early in U.S. usage of cell phones, the network effect was limited since there were relatively few people whom you could call and reach on their cell phones. Over time, the number of people who keep their cell phones on has increased, thereby increasing the value of the cell phone network.

Another important change in style of usage over time has been the placing of cell phone calls to "stay in touch" and socialize. One example is a person who calls home from a train or airport not to report that they are going to be late but to say that there is nothing to report and that they are on time. Earlier, when cell phone service was expensive, usage was more deliberate—people called others from their cell phone because they had a specific message to convey. As the price dropped, people began to call others just to keep in touch, often with no specific message. This has helped the cell phone to become a regular habit for millions of users, sometimes to the annoyance of those nearby in public places who overhear constant chattering on cell phones. This pattern also parallels the early and later uses of wired telephone lines. In the late 19th- and early 20th-centuries, telephones were used primarily by businesses and wealthy people to convey specific messages; later, they were used by the mass public to convey messages and also to chit-chat or socialize (Fischer, 1992).

It is also noteworthy that people have accepted a lack of privacy in cell phone use and engage in behavior that would have been deemed bizarre only a decade ago. On trains, in public buildings and in the street, they use cell phones for personal conversations that are easily overheard by others. The enclosed telephone booth has seemingly been abandoned as a way to maintain privacy while making a telephone call outside the home or office.

As the cell phone has become used more and more, some people have reduced their use of wired telephones and think of the cell phone as their main phone. Others have given up wired telephones at home (Tahminclioglu, 2004). It appears that these are largely younger people who have grown up with cell phones and use them in the household as well as in mobile situations. Often, they purchase a cell phone plan with thousands of minutes of use per month, so it is cheaper to rely on the cell phone exclusively and forego a wired phone line.

2.2 Short Messaging Service

Short Messaging Service (SMS) or text messaging is an abbreviated form of e-mail. It evolved from attempts to display e-mail on cell phones and other devices with very small screens. Long messages are very difficult to read on small screens and even tougher to type on the tiny keys of a cell phone or virtual keys on a PDA. Many SMS services limit messages to 150-160 characters or approximately 25 words. Further, they may cut off long messages sent through regular e-mail. In practice, most SMS messages are much shorter than 150-160 characters. Further, they use various abbreviations to reduce typing time and make more efficient use of limited screen space, for example "G2G" for "Got to go" or "WGD" for "What are you going to do?" SMS has also led to a new form of typing, sometimes called "thumb typing," in which people type with one or two thumbs. Many teenagers have become quite adept at thumb typing. A number of older users then followed the lead of the younger generation.

SMS is the most successful early wireless data service among the general public, particularly in Europe and Asia (see the discussion of wireless data services across cultures below). It is less common in the U.S., but had been widely adopted by users of the RIM Blackberry service. Further, many teenagers in the U.S. use SMS on their cell phones. SMS has also inspired a new form of advertising—short ads sent as SMS messages to cell phones and wireless PDAs. Early research indicates that the response rates to SMS ads are high, but it is unclear if this high response rate will continue: banner ads on the Web had high initial response rates, but they dropped over time.

2.3 Wireless Data Services Across Cultures

Among the general public, wireless data services have been more broadly adopted in Europe and Asia than in the U.S. However, a number of proprietary business applications of wireless data services have become commonplace within the U.S. and consumer adoption is catching up.

In Europe, SMS has been widely adopted in a number of countries and has generated significant revenue. Throughout Europe, approximately two thirds of the 450 million mobile phone subscribers use SMS. The SMS phenomenon began as a teenage and 20-something craze, then spread to older demographic groups.

On a smaller scale, there have been some trials and limited services in Europe that use the cell phone as a payment device for transactions at vending machines, car washes and fast food restaurants. For example, Sonera Corporation has developed applications in Finland and Sweden. In these applications, a cell phone user dials a special number displayed on a vending machine or service counter and authorizes a charge to their credit card for the soda, fast food, or service, which is then dispensed by a vending machine or provided by a merchant. It remains unclear whether these small transactions will be a viable business model for wireless devices and if this form of payment will be widely adopted. In addition, there has been modest use of premium content services such as paying for special ringer tones, sports scores, and stock prices.

In Asia, SMS is also common among cell phone users. Indeed, many cell phone users in Asia prefer SMS to e-mail . Much has been reported about wireless data services in Japan, notably the DoCoMo i-Mode service. Most of these services are inexpensive: $2.50 or less per month (some are less than $1). The fee is automatically added to the subscriber's cell phone bill. What are people paying for? Among the most popular services are customized ringer tones, virtual pets and animated characters downloaded to the cell phone. I-Mode also provides access to a few thousand Web sites that have been reformatted for i-Mode. Some of these are free and some are premium services. Reformatting for wireless data services is not easy or cheap. Nikkei, Japan's largest business daily, has a "Short Message Team" that designs software to automatically trim content for i-Mode and two competitive wireless data services (Shimbun, 2001). It is crucial that content in a long-form medium such as a newspaper or magazine be adapted to the requirements of small screen displays.

Wireless data activity in Japan is very strong in part because of the high penetration of cell phones. Japan has more cell phone subscribers than wired line subscribers: 86.6 million cell phone subscribers versus 71.1 million wired line subscribers in 2003 (ITU, 2004). Japan's commercial infrastructure also lends itself to potential wireless data services. For example, Japan has an extraordinary infrastructure of vending machines—many more than in the U.S. or Europe. Over $55 billion in products are sold from vending machines in Japan each year (Yamada, 2001). This is an attractive base for developing wireless credit card applications. In addition, the i-Mode service is always on, which lends itself to location-based services such as tracking the movement of vehicles. As a result of all these activities, the wireless data industry has followed Japan very closely, looking for applications that are successful and which might be duplicated elsewhere.

In the U.S., the use of SMS was initially hampered by a lack of interoperability among wireless services but has grown in popularity, especially among teenagers. SMS has become very popular in the U.S. among a group of largely corporate users of the RIM Blackberry, which shares the 'always on' feature of i-Mode and is easier to compose messages on compared to cell phones. However, its relatively high cost ($40 to $60 per month for wireless data service) has restricted adoption to corporate users and some consumer aficionados of wireless data services.

There have also been some tests of mobile commerce in the U.S., using PDAs with wireless modems as well as cell phones and some attempts to set up wireless local area networks in coffee shops and airports where users with specially equipped laptops that can access the Web at high speeds. However, these early efforts have not yet turned into viable businesses. Nonetheless, interest remains high among service providers for these high-speed wireless local area networks.

Lost in much of the discussion about U.S. wireless data services are the many successful proprietary applications of mobile data services in business settings. These include Federal Express, UPS and many other businesses that equip delivery, sales, repair, and other personnel with wireless data terminals that can send and receive information. These applications track deliveries, record and transmit sales, provide access to inventory information and manage the movement of personnel in mobile settings. While there has been relatively little attention given to these wireless data successes, there may be important clues in these services for the development of next generation wireless applications.

3 Assessing Demand for Next Generation Wireless Services

Next generation wireless networks represent a broad array of end-user devices, network configurations and speeds. Service applications must take these differences into account. For example, will the end-user device be a cell phone, PDA, laptop, or hybrid device? Each of these devices has different capabilities. Some wireless networks reach across wide regions; others are clustered within a few hundred feet of a coffee shop, airport, office building, or other locations. In the latter case, a person must know where the network is located and have the necessary equipment and account to tap into it. The tradeoff is that the clustered networks such as WiFi can have very high transmission speeds (up to 11 Mbps), while the broad regional networks will likely offer services at somewhat slower speeds. To add to the complexity, the actual speed that an end user experiences (and which a service provider must plan for) can vary considerably from advertised speeds. Planners must decide whether they can create a service for one device that will be used on one type of network at a relatively constant speed or if the service must support multiple devices, multiple speeds, and many types of locations. The latter places considerable constraints on the planning of new services. End users face a similar set of issues in deciding what device they will need, how it must be configured, where they can access services, and how those services will perform at various access speeds. Further, all of these parameters will change over time. Next generation wireless services are a moving target.

3.1 Core Issues in Assessing Demand

There are a few core issues that are relevant to an assessment of demand for most new technologies, including wireless networks. The first is whether next generation wireless networks will provide a host of new services or (primarily) enhancements to existing services? Many new technologies have thrived by providing desired enhancements to what already exists. Color television and graphical user interfaces are two examples. They made the experience of technologies that already existed (B&W TV and text-based computing) much more enjoyable and user friendly, leading to rapid adoption of a new generation of the technology. In the case of wireless services, desired enhancements might be as simple as greater reliability, improved voice quality, and quicker access to data services such as messaging.

Alternatively, widespread adoption of next generation wireless services may require some new applications. If so, what new applications? This leads to the perennial search for the 'Holy Grail' of new technologies—killer applications that are so highly desired, that many people will pay significantly increased fees for the service. Unfortunately, it is rare to find killer applications that are genuinely new. Communication appears to be the most important killer application for wireless, both in terms of voice and data services. However, it is not new. Among content offerings, pornography, gambling, games, and shopping have been killer applications for many earlier generations of technologies such as the Web. None of these is really new. More commonly, it is a package of many service offerings, with a few standouts such as messaging, that attracts new users and converts existing customers of the earlier generation of technology.

It is also important to distinguish the reasons why people acquire a new technology or service and how they actually use it. This is particularly true for the earliest adopters who often acquire a technology in order to be able to brag about it with friends or for some exotic feature that they then use occasionally or rarely, e.g., handwriting recognition on PDAs. Similarly, a larger group of users may buy a technology for a core value, for example, women who acquired cell phones for "security," but then use it primarily for other functions such as coordinating schedules or socializing. This is not to argue that exotic features or highly-valued but rarely used features are not important. They are significant marketing tools. However, a service that has continuous monthly fees (versus a product that is purchased and then has no ongoing fees, such as a PDA) must give people a reason to pay each month for the service. Further, business organizations may be surprised to find that some salespeople demand next generation wireless services for access to inventory databases but then use the devices primarily for games.

When new service offerings are driven largely by advances in technology, it is necessary to ask if this is just technology being pushed at the marketplace or if there is an existing or latent demand? This question is often used to critique new technologies. However, a historical perspective on the introduction of many new technologies, for example, telephones, radio, TV and personal computers suggests that there was no existing demand for any of them. It may be argued that there was a latent demand but it appears more accurate to say that they created a demand. This may be the case with next generation wireless services. There does not appear to be a strong existing demand for these new networks. There may be a latent demand, especially to fix problems with current wireless networks. The challenge is to create demand by demonstrating services that people will want once they see what these services can do.

Two core questions facing developers of next generation wireless devices are: which devices will people favor; and will end users prefer one device that does many things or multiple devices, each of which does one thing primarily? The candidate receiving devices for next generation wireless services include cell phones, PDAs, pocket PCs, laptop computers, pagers, proprietary devices created for a specific application or user group (the Fedex wireless data terminal), or a hybrid of one of these. The shape and weight of some devices have limitations for certain applications, for example, current PDAs are an odd shape to hold to the ear like a cell phone and require a supplementary earpiece and microphone for making telephone calls; a laptop weighs a lot and is not as easy to carry around as a PDA or cell phone. Does an all-in-one device make sense? In some of the author's research, many people initially liked the idea of an all-in-one device for next generation wireless services, until they realized that their all-in-one device built into a cell phone would add considerable weight and size, and that a laptop wouldn't be a convenient portable phone. This brought them back toward a desire for multiple devices, but hopefully two devices and not five. Adding to the challenge for designers is the obstacle of adding many features to a device such as a PDA or cell phone without creating usability problems due to complex interfaces that guide people to multiple services within the menu structure.

3.2 Applications

There is a broad range of applications under consideration for next generation wireless services. They include consumer applications such as messaging, mobile commerce, entertainment, videophones, medical applications, advertising, and location-based services; business applications such as management of workers in the field, inventory access, sales placement, remote sensing and monitoring, and shipment tracking; and government or military applications such as battlefield command, crime monitoring, and disaster recovery and assistance. A comprehensive assessment of all of these applications is not possible here. The discussion below emphasizes key elements that are likely to affect the success or failure of several prominent applications.

Messaging in the form of SMS or longer e-mails that will be possible on next generation devices will build upon the broad acceptance of text messaging for mobile devices in Europe and Asia, and by many in the U.S. Voice messaging services such as automated voice reading of a sent e-mail, voicemail and providing audiotex services such as sports scores, stock quotes, and horoscopes have received less attention than text messaging but in the past, these have been popular.

Much attention has been given to mobile commerce services (m-commerce) such as buying train or movie theatre tickets, paying for fast food orders, checking-in at airports, and purchasing small items such as a can of Coke from a vending machine. Clearly, the carrot of a "wireless credit card" has enormous appeal for wireless network developers and there are many potential advantages for consumers. Think of two movie theatre lines: one in which people have to purchase a ticket with cash or a credit card; and one in which a person can pay with cell phone or enhanced PDA and walk directly into the movie house. For merchants, the carrot is in fewer sales clerks, ticket agents, etc. However, the challenge will be in the implementation. For m-commerce to be successful, there will be a need for broad interoperability, wide adoption in the marketplace, a business model that works for small transactions, and transaction speeds that are faster than face-to-face transactions.

Entertainment is also high on the list of potential applications for next generation wireless services. Some envisage on-demand movies and live video over 3G networks (*The Economist*, 2004). However, this appears to be stretching the limit of what 3G can deliver. Further, the viewing experience on a two or three-inch screen may be less than appealing. Also, it appears to be an attempt to reinvent broadcast television, which does a much better job of delivering high quality video. Nonetheless, entertainment services, especially games, have received moderate usage in Europe and Asia (Dvorak, 2004). Next generation wireless networks could provide a platform for delivering higher-end games. In addition, audio entertainment is more feasible on 2G or 3G networks. A number of services offer music, turning a cell phone into an MP-3 player.

Location-based services would combine the features of 3G and global positioning system (GPS) to locate where a person is and then provide information such as a list of nearby restaurants or directions on how to get to grandma's house from wherever you happen to be. It has practical uses and also a 'cool' factor that is likely to appeal to males who do not like to stop at the gas station and ask for directions.

Advertising is everywhere, including plans for next generation wireless services. Ads will be necessary in services that are offered free, but it is hard to gauge the annoyance factor when screen size is so limited. A banner ad on a 17-inch PC Web screen can be relatively unobtrusive. It is hard to imagine any ad on a three-inch screen that would not be perceived as obtrusive. SMS ads are less obtrusive unless they are so numerous that they clog up menus of incoming e-mail. However, there may be clever alternatives such as electronic coupons or advertising information on request that people will accept. Early reports about high response rates to wireless ads need to be viewed in the context of early usage and early adopters. Response rates are likely to drop over time, as they did with banner ads.

There are many potential medical applications for next generation wireless services to provide greater mobility for people with significant medical problems. Many of these involve monitoring a patient's vital signs and communicating them in real time to monitoring stations at a hospital or other facility. Other consumer applications range from the curious to the far-fetched and include using a 3G device as a hotel key, placing electronic tags on children that use GPS to communicate back to a parent's PDA and tell the parent where the child is, videophones, and dating services that keep profiles of members and light up cell phones or other devices when two compatible people walk near each other. It is difficult to judge which if any of these services might find widespread acceptance. However, it is important to understand the history of similar devices. The videophone, for example, has been tried several times and has yet to find consumer acceptance. New videophone services may experience a different fate but they will have to overcome well-known problems such as poor image quality, feelings of embarrassment by some people in being seen and difficulty in maintaining a consistent orientation toward the camera—it is very easy to inadvertently send someone a videophone image of your tie instead of your face. Still image transmission is technologically more feasible and has proven popular in Japan.

Enterprise applications may offer greater near-term opportunities than consumer applications, at least in the U.S. The list of potential applications is long and includes supply chain management, emergency communications, customer relationship management systems, maps on demand, re-routing trucks, managing the schedules of mobile workers, operating a completely mobile office, and providing assistance to mobile workers, among other applications. There are also many applications for specific professions such as the military and journalists. These groups have greater financial resources for applications that can demonstrate genuine utility and they can provide technical support and training that may be necessary for some applications. They also have a number of concerns. In research by the author, concerns about next generation wireless services by business users included battery life, weight, readability of screens, size of the device, and actual versus advertised speed of access.

4 Important Features and Design Challenges for Next Generation Wireless Services

One feature that is present in some current wireless services such as i-Mode and Blackberry, and which is planned in most next generation services, is 'always on.' That is, the device is always turned on to receive incoming voice and data services. Further, these can be sent to the device without any action by a user. 'Always on' appears to have had a significant impact on usage. Further, in Europe, where cell phone users do not pay for incoming calls, the billing structure encouraged people to keep their cell phone on and thereby increased the overall value of the cell phone network for users. In the U.S., where cell phone users pay for incoming calls, the billing structure initially discouraged many people from keeping their cell phones on unless they were making a call (this changed, when prices dropped low enough for people to encourage incoming calls). An 'always on' state encourages more usage, speeds up the transmission of messages (since they can be received at any time) and supports real-time information services which also add value to the service (e.g., it is more valuable to receive a quote about a stock that has dropped in price as soon as the event occurs compared to an hour later).

Screen size, shape, color and resolution are also critical issues for next generation wireless services. There are a series of tradeoffs associated with these features. On the one hand, a larger screen with higher resolution and color will enhance the readability and visual appeal of services. On the other hand, these features will add cost and/or weight to a device. From a service provider's perspective, a major challenge will be to design services that are usable on a broad range of screen sizes, resolutions, color versus black and white, and screen shapes. It is unclear whether content can be adapted automatically to "fit" onto many different screens or if human designers must intervene. The latter will add significantly to the cost of providing services.

The user interface and navigation for next generation wireless devices also present a challenge. First, navigation options are limited on small screens. This can lead to multiple layers of menus, which users consistently dislike. Further, important features and services are often lost in the confusion and sheer work of navigating through multiple menus. Icons are often used to save space on menus but research on icons indicates that users are more confused by icons than helped (Kansas, 1993). User interface and navigation challenges for small screens can be solved. However, it is a costly and time-consuming process that many service providers do not value, or, put off until the last step in developing a new service, at which point options have been restricted because of software and hardware decisions that have already been made. Further, in the rush to get products to market, many product developers simply cannot afford the time to design and test user interfaces in a thorough way. We see the consequences of these inadequate efforts in many electronic products that are introduced into the marketplace.

Text entry and selection commands are also vital to the success of next generation wireless devices. As devices shrink in size, it is less feasible to use full size keyboards or a mouse to enter text and make selections. However, there has been considerable progress in this area, with better designs for micro-keyboards, thumb-controlled pads that replace a mouse, and voice recognition technology that works for many simple selections and commands. However, the proliferation of devices with many different text entry and selection features can create a burden for service providers who seek to support all of these devices and the many ways in which they organize and label choices. Imagine creating software for a PC universe in which there were a dozen different keyboards, each with different labels for Backspace, Enter, and Shift and some of which replaced the Enter key with three separate keys, each of which was used for different types of Enter commands. Interoperability involves not just the ability of one network to talk to another but also for software and wireless services to support many alternative devices with different features.

All of these design challenges to creating functional and user-friendly devices must not get in the way of providing a feature that will sell the next generation hardware to many early adopters, that is, a "cool look and feel." An "icy steel blue" body, a color coordinated faceplate, a shape that "feels good" in the hand, and other design features that have little to do with functionality often sell millions of devices. Technology reviewers for magazines and newspapers, as well as consumer and business users, often pay as much attention to perceived design 'coolness' as functionality. However, if coolness sells, good functionality is why a person continues to use a device a year later.

5 Discussion

A critic of next generation wireless services could argue that if one eliminates videophones, movies-on-demand and a few additional high-end services that appear to have limited appeal in a 3G wireless environment (or simply will not work very well even in a 3G environment), all of the other services discussed in this chapter could be offered with the current generation of wireless networks. Even in Europe and Japan, robust wireless services for a mass market require more bandwidth. How much extra bandwidth is the crucial question.

This review of research and marketplace experiences does support the argument that the most feasible path to next generation wireless services is likely to be enhancements to current services that consumers and businesses want and use. That is, an evolution of services rather than a radical break from current to completely new services. Further, the next generation of wireless networks will provide an opportunity to fix problems with current services and, generally, to learn from experiences to date. This path of development is not dissimilar to what has taken place with broadband Web services. Although broadband Web networks have provided access to some new services, most people have adopted broadband for faster access to services they were already using. Further, they are willing to pay for this faster access.

The review of research in this chapter also suggests that the best way to understand demand for new services is to focus on the core functionality of a wireless network and the attributes of the mobile society in which we live. Specific applications and services can follow on this understanding of what wireless networks mean in people's lives and what are the broad needs of a mobile society.

5.1 Differences in Adoption Across Cultures

There are a number of differences in the adoption and use of current wireless networks across countries in Europe, North America and Asia. Differences are likely to continue with the next generation of wireless technology. Some of these differences relate to technology and regulatory structure; others to cultural differences. For example, Europe got started with the development of 3G networks (i.e., by assigning spectrum through license auctions) before the U.S. However, some argue that the U.S. is now in a good position to not only catch up with but avoid some of the mistakes made in Europe, for example, paying too much for 3G spectrum (Parker, 2004). In terms of infrastructure, the U.S. has more personal computers and Web access but fewer cell phones per capita than many European countries and Japan. So, the U.S. may be more inclined to use wired Web access for some applications, where Europe and Japan might turn to wireless networks. The differences, however, are narrowing.

Regulatory environments also differ and will affect how next generation applications are developed. For example, the U.S. has a more relaxed regulatory environment compared to Europe, so the U.S. encourages market forces to work out issues such as standards and is generally entrepreneurial in allowing multiple services to compete. Europe believes more in central planning and the adoption of single standards. The European regulatory model was clearly advantageous in supporting the fast rollout of 1G services. Will the same hold true for 3G?

Cultural differences are the most intriguing and probably the least understood in terms of impact on next generation wireless service development. Will Americans pay as much for premium content and customized services, as have millions of Japanese? What has been the role of off-color or slightly pornographic SMS messages in Europe and will these types of messages carry over to next generation services? What about the American penchant for gambling, for example, billions of dollars bet on major sports events? Will organizations exploit these habits in next generation wireless services? What about personal space and the need for privacy? It is reported that one reason Japanese teenage girls have adopted SMS is that it affords them privacy in households with limited personal space. What about evolving etiquette issues such as the acceptability of people sending SMS during business meetings (Richtel, 2004)? These and many other cultural factors are likely to influence the adoption of next generation services in ways that few understand.

5.2 Pricing and Adoption of Next Generation Services

This chapter does not address the issue of business models or the economic prospects for next generation wireless services in the broad context of costs to acquire spectrum, build networks and develop services. The issue under discussion here is simpler: how should pricing for new services be structured and are people likely to adopt next generation wireless services that cost more than their current wireless service? Japan and Europe have demonstrated that people are willing to pay more for enhanced wireless services such as SMS and some content services. In the U.S., people have shown a willingness to pay for higher tiers of wireless service that include extra minutes, no roaming charges and value added services such as caller ID. However, in all of these cases, the fees paid have been modest: $1-$3 per month for premium content services in Japan; 10p per SMS in the UK, and $10 to $30 per month for higher tiers of service in the U.S. The important point, which should not be overlooked, is that many people have shown a willingness to pay for extra services. By contrast, the wired Web has become an economic trap for many service providers who encountered an attitude of "We don't pay for content" from users. The Web provides another lesson: there is a danger from a service provider's perspective in giving away content for free in the hope that people can later be converted to paying subscribers.

Moving people up to next generation services may accompany replacement cycles for existing cell phones and cell phone contracts. In the U.S., the average cell phone is replaced every 14 months; the average laptop computer every three years. If next generation wireless services are viewed as an enhancement to current service, then the replacement cycle for the receiving device might be the decision point for change. Under this optimistic scenario, next generation wireless services might follow a path similar to audio cassettes that replaced long playing records and CDs which replaced audio cassettes. However, the picture is likely to be more complex. If the pricing of next generation services is high, there will likely be a need to find early groups who are willing to pay higher prices for these services. In the U.S., business users are an obvious target group and some service providers have already begun to target them. The same may hold in Europe and Asia, but in those markets some consumers have already shown a willingness to pay for enhanced wireless services, although spending on these services has been modest and based upon inexpensive per use fees (in the case of SMS) or relatively low monthly fees for content services.

5.3 Content Models

There are many unanswered questions about the best way to structure services and content. One obvious question is whether some or all content should be free, supported by advertising, or paid for directly by end users. The advertising model is questionable given the screen size limitations of most planned wireless devices. How can ads be presented in a way that is appealing to the advertiser but not obtrusive for the end user? Further, it may be difficult to convert end users from free to paid content, if the advertising model is not successful.

A second question is whether there is a need for robust content offerings, for example, make the entire Web available through next generation wireless services, or offer limited content that is designed for small screens and whose quality can be controlled. The former will appeal to many potential users but the experience of trying to navigate and read voluminous content within a small screen environment may be disappointing. Web content must be reformatted for small screen access and viewing. The latter approach, which is sometimes called a 'walled garden' model, can increase the chances of a positive content experience but it must be presented very carefully so that a user's expectations are set appropriately and met successfully. The danger is that walled gardens might be perceived as weak and limited service offerings.

5.4 Open Networks and Creativity

There are many other applications under development for next generation wireless services. Some of these have been discussed briefly in this chapter, for example, wireless credit cards, emergency services, location-based services, video downloads, MP-3 phones and networked games, among others (Park, 2004). No one can predict which will be successful and which will not. The question is: Will they be allowed to test the marketplace? Next generation wireless network operators can create an environment where creativity flourishes and many groups can develop services or they can build a controlled environment which limits access. The history of other media, for example, the telephone, radio, television and the Web, demonstrates that early builders of technology infrastructures often have a myopic view of the potential for the technology. Creative ideas and, typically, successful models of content and services, are often discovered after the technology has entered the marketplace. Further, they are discovered by creative entrepreneurs who are not technologists and who are outside the technology organizations that created the medium.

References

Dvorak, P. (April 23, 2004). Big jump in games in Japan among teens. *The Wall Street Journal*, B1.

Dvorak, P. (June 3, 2004). Testing the TV tuners and fingerprint checks in cellphones in Japan. *The Wall Street Journal*, B1.

International Telecommunication Union. (2004). *World Telecommunication Indicators*. Geneva: ITU.

Kansas, D. (November 17, 1993). The Icon Crisis: Tiny Pictures Cause Confusion. *The Wall Street Journal* (p. B1).

Katz, J. (1999). *Connections: Social and Cultural Studies of the Telephone in American Life*. New Brunswick, NJ: Transaction Publishers.

Parker, G. (August 12, 2004). Race to link WiFi, cellphones pick up. *The Wall Street Journal*, B4.

Richtel, M. (June 26, 2004). For liars and loafers, cellphones offer an alibi. *The Wall Street Journal*, C1.

Shimbun, N. (April 9, 2001). The Web @ Work. *The Wall Street Journal* (p. B4).

Tahminclioglu, E. (February 22, 2004). Cutting the home cord, not the home number. *The New York Times*, 9.

Yamada, M. (April 16, 2001). Press 1 for coke or # for condom. *The Industry Standard* (p. 26).

III

Business Models

10
Profitable at any Speed?

Bertil Thorngren

1 Introduction

Some years ago the promise of i-Mode in Japan, Blackberry in the U.S., and Short Message Services (SMS) in Europe aroused interest among investors worldwide. The reasoning seemed to be that, if slow-speed services such as these were so much in demand, then increasing the speed and capacity in order to provide even richer content would uncover new demand and new revenues to be shared among operators, content providers, and other vendors.

2 The Technologies

Some of this might actually come true for the operators of Wireless LANs, also known as WLANs, WiFi, or 802.11, now rapidly being deployed at airports, railway stations, hotels, restaurants and other public spots, and which can even be made available inside aircraft and trains. As these services can provide up to 11 Mbps or even 54 Mbps they have a claim for the much-touted marriage between wireless and the Internet. However, the very merit of providing "true broadband" takes away a bit of the newness seen from the perspective of the content providers. Pricing and other aspects of the business model might look pretty much like those that apply to fixed broadband and other flat-rate services. Some possible exceptions to this simple observation will be discussed later in this chapter, after new mobile services such as 3G and 2.5G have been brought into the picture.

Cell phone operators have settled into a "generation" mindset, where 1G analog voice-only networks were replaced by 2G digital voice-only networks, which in turn were to be replaced by 3G voice and data networks. The movement from 1G to 2G went well, but the jump to 3G has not.

The original concept of UMTS (3G) was laudable especially since users were assumed to be able to use the same device in Europe, the U.S. and in Asia. That vision has foundered on the rocks of a lack of coordinated spectrum between the U.S. and the rest of the world, spectrum scarcity most everywhere, resultant high prices for the spectrum available, hardware and software development problems, and perhaps most importantly, customer disinterest.

The 'solution' has been a 'mid-generation' or 2.5G solution (also known as GPRS), offering low-speed data and digital voice in the existing 2G bands. I-Mode, SMS, Blackberry, and most other cellular data service currently available live in this generation.

The outlook for mobile operators geared to provide true "3G" services looks far from hopeful. Providing higher speeds implies not only higher costs, but also less revenue per MHz compared to using the available and scarce spectrum for less "capacity-hungry" applications such as voice, SMS and email. Consider the following:

Each 3G operator has a finite number of channels to offer customers per serving point and, although the channel may be used for either voice or data under 3G, it cannot be used for both at the same time. Therefore, any data channel in use takes away one voice channel. Since the 3G model calls for data services to always be available, the ability of a serving point to service voice calls will be reduced by the number of 3G data users on a one-to-one basis. Now consider the pricing model for data use throughout most of the world. Users expect to pay for packets or for flat-rate access, not time of connection, and neither volume nor flat-rate pricing is likely to replace lost revenue from voice or SMS-type usage.

Even sending and receiving digitized photos directly through a cell-phone or a PDA does not require more bandwidth compared to what is now used for voice over mobile networks or faxes over fixed as well as mobile networks. Thus, Multimedia Message Services (MMS), which is supposed to supersede simple SMS messaging can well be handled even at present speed levels of 9,6 kbps, especially if supported by Java and other means to minimize the need for more transmission capacity.

As indicated above, the proven success stories like i-Mode in Japan, SMS in Europe and Blackberry in the U.S. are working at low transmission speeds. However, they are offering other qualities that might be of greater importance to users and customers, like reliable nationwide coverage and low cost for each transaction, not to speak of the importance of "always on", an inherent feature of packet switching as opposed to traditional circuit switching. Getting a laptop up to work is still a matter of minutes rather than seconds. Any device and service that can provide instant access is a winner for those on the move. This might well be one of the reasons behind the success of i-Mode in

Japan as opposed to the failure of WAP in Europe that was launched prematurely, before the availability of packet switched services.

Given these customer reactions, voting with their feet and their wallets, how is it that the mobile operators have been so obsessed by the increase of sheer transmission speeds rather than other qualities where they can still claim a unique advantage? Other options like WLANs can offer radically higher speeds (Mbps rather Kbps) at a rapidly increasing number of Hot Spots, but nationwide coverage and roaming is still far off.

Anywhere/Anytime/Any Device on Anybody's Network is a truly forceful concept, even at more modest speeds, especially if access by traditional mobile services can be seamlessly combined with the higher speeds provided over WLANs. From a technical perspective it is already quite feasible to offer uninterrupted sessions over networks running at different speeds (Thorngren et al., 2004).

Commercial roaming agreements are quite another matter, as mobile operators are still pondering over whether WLANs are to be seen as friend or foe (Lehr & McKnight, 2005). The initial knee-jerk reaction of mobile operators in France and the UK was to ban public WLANs, considered to be a threat to the revenues much needed to pay for extravagant 3G license fees. WLANs might lower their (projected) revenues by tens of percentage points. However, these projected revenues look to be very problematic, even without taking WLANs into account. The operators have rather good reasons to welcome public WLANs as a blessing in disguise rather than any threat to any theoretical revenues.

As shown above, any migration to higher speeds in a mobile network is bound to provide less revenue per MHz or Mbyte compared to plain voice or other not so capacity-hungry non-voice services, like SMS and transmission of still pictures. Any downloading of hi-fi music, not to speak of movies at higher speeds than 100 kbps, is simply not realistic unless the price per Mbyte is drastically lowered to a level encouraging arbitraged delivery of more basic services. The rapid deployment of more cost-effective Wireless Internet Service Providers (WISPs) only underlines this fact of life.

3 Implications for Content Providers

Classical vertical integration is being replaced slowly but steadily by more horizontal market structures. Content providers are not bound by the treacherous concept of "generations" (like 3G) which supposedly deliver radically new opportunities. Digital content can readily be re-packaged for access through a plethora of different networks and devices. On the face of it, mobile services provide another opportunity to get paid for content, repackaged to differentiate it from that available over flat-rate Internet services (Funk, 2004).

Revenue sharing between content providers and mobile operators has already become more common. However, there is simply more revenue to be shared by providing high-interest but low-bitrate content, like ring signals, logos and e-mail services, compared to more capacity-hungry applications. A possible way out from what looks like a Catch-22 for content providers and mobile operators alike could be a multi-channel approach. Operators of "3G" networks could prosper by providing anywhere/anytime instant access, coverage and position-based services, including giving "always on" referrals to locations where more content-rich services, provided by WISPs, can be accessed and downloaded.

In any case mobile operators look bound to charge for capacity used, whether measured by minutes or Mbytes, given the scarcity of frequencies. By contrast, most providers of public WLANs look bound to provide flat-rate services, even if access can be conditioned by permitting use only per day or per hour. Others, like airlines or hotel chains, might find it of interest to provide free access (at least to their own websites) in order to attract new customers.

Perhaps a bit more discussion of "value-based pricing" is in order. "Value-based pricing" is already in full use when it comes to basic SMS. A fixed price is natural because the content size is also fixed (to 160 characters, total). When it comes to "premium SMS" with logos, ring tones, and such pricing on "perceived value" is also natural because the revenue is to be shared between the content provider and the operator. Fixed pricing is also logical when it comes to sending MMS-messages such as still pictures, which again have a fixed maximum size and use only a fraction of the capacity needed for voice calls.

There is, however, a limit to how far this fixed price (for a given content) can be scaled up. For capacity-thirsty applications, like downloading music or films, the sheer capacity cost increases in proportion, taking up more and more of total "perceived value" and leaving less and less room for a content provider to get its share. To some extent this can be met by differentiating the offers, such as a lower price asked for a low-resolution and brief streaming video version, and a higher price for the full version. As stated elsewhere, any full video version over a cell phone might however simply be too expensive to attract more than a handful of customers, as customers have the choice of some other and far less costly option, like a WLAN, for any serious downloading. In the early days of 3G some operators might well choose to offer almost unlimited usage at a flat rate. This is not a sustainable strategy however as any 3G network has only so much of total capacity.

When it comes to customers browsing the website of a newspaper, it is up to the customer to decide how long any session might last, as well as the amount of downloading required, which might necessitate some kind of billing measured in minutes and/or Mbytes.

The unregulated WISP/WLAN business has unresolved challenges of its own. Some services might be provided for free, sponsored by airlines and hotel chains, banks and so on, while other services might be more costly as the owners of the most attractive venues see a new opportunity to exploit the value of its location and hence a captive market.

From the perspective of content providers this is a new challenge. Some might prosper by providing "must have" content. Others might suffer if they either fail to generate interest or subsidies from airlines, hotel chains, and so on, or if venue owners ask for too large a share from a yet undeveloped WLAN business opportunity.

In any case, the market for wireless is not only a matter of sheer content delivery. Quite a large share of the market might simply be produced by the end-users and customers themselves, not simply by email but photos of the kids sent to grandparents and photos from tourists sent back home rather than postcards. Within the business sector, access to the company Intranet (more or less internally produced) might be the driving force. The human urge to communicate is fact of life, which even Graham Bell failed to recognize. He assumed that the telephone was to be mainly used for access to concerts rather than peer-to-peer communication. That said, there might be a business opportunity in the editing or processing of customer-produced content.

4 In Summary

An increase in speed will not necessarily enhance revenues and profits, but rather create the contrary for operators of mobile networks and content providers. Other aspects than sheer speed are more likely to attract interest and willingness to pay. Multi-channel approaches, combining the virtues of always on/anytime/anywhere and far higher speeds/lower costs provided by WLANs look like a winning combination. There is no need to look into the future for a "4G" solution, because it is actually an option available today.

From the perspective of content providers, 3G operators (and 2.5G operators as well) might be of interest as discussion partners. Even if operators might be reluctant to put any high-priced bills for content on their own invoices (already high enough), they might be able to offer other convenient models for billing and revenue sharing. Content providers need not even consider services requiring more than around 100 kbps, as higher data rates are unlikely to be delivered by 2.5G or 3G at an affordable price.

Delivery via WLANs is clearly more interesting when it comes to capacity-hungry applications, like downloading of music and movies or large business documents or presentations at off-site locations.

There is, however, a possible combination of the two worlds, which an increasing number of operators worldwide has already embarked on.

Hopefully this new combination will offer users "Seamless Mobility" as well as an opportunity for the content industry to more freely chose and combine alternate distribution channels.

References

Funk, J. (2004). *Mobile disruption. The technologies and applications driving the mobile Internet.* New York: John Wiley & Sons.

Lehr, W. & McKnight, L. (2005). Wireless Internet access: 3G vs. Wifi? In ECC (Ed.), *E-merging media. Communication and the media economy of the future* (pp. 165-180). Berlin: Springer.

Thorngren, B., Andersson, P., Bohlin, E., & Boman, M. (2004). Seamless mobility: more than it seems. *Info*, Vol. 6, Number 3, 169-171.

11
Mobile Commerce Business Models and Network Formation

Carleen F. Maitland

1 Introduction

New generations of wireless and mobile communications networks stand ready to revolutionize the global media industry by creating a ubiquitous and personalized channel to consumers. The extent to which this revolution will occur however will depend on a variety of factors, ranging from technical performance (Dehghan et al., 2000) to user acceptance (Anckar & D'Incau, 2002). Despite these uncertainties, the rapid increase in Japanese subscribers to mobile information services has fueled optimistic expectations elsewhere. Riding this wave of optimism, firms hoping to get involved will quickly see it is nearly impossible to 'go it alone.' An examination of the business models for currently available mobile content and information services, such as weather forecasts, banking services, and online gaming, reveals that coordination of a wide variety of firms is often required. Increasingly, this coordination occurs through a network of firms that comes together to provide a service. Network formation for service provision may help spread the risk of developing new services, however it may also increase the challenge of achieving financial success.

To some extent however these challenges are not new. Mobile information and entertainment services join a growing list of services from a variety of industries, ranging from automotive goods to biotechnology, which are produced through increasingly complex networks of firms (Hage & Alter, 1997). This trend was also evident in business models for electronic commerce, from which mobile commerce has much to learn. There are nevertheless a number of differences between the two, such as billing capabilities, greater levels of personalization, and network access, which will influence mobile commerce business models. At first these expanded capabilities may appear to broaden the range of possible business models, however this propensity will be tempered by the consensus that working in a network of firms requires.

The aim of this chapter is to investigate the issues inherent in mobile commerce business models that are developed and implemented through networks of firms. The investigation begins with a discussion of business models and the unique aspects of mobile commerce. This is followed by an exploration of the factors driving network formation in this industry. Subsequently the issue of power in inter-organizational networks and its effect on mobile business models are described. Next, using examples of the business models of two mobile commerce firms, the implications of network relations for business models are discussed after which the chapter concludes with suggestions for future research.

2 Business Models and Network Formation for Mobile Commerce

2.1 Business Models for Mobile Commerce

A business model is one of many tools used by a firm to develop new products or services or to revise existing offerings. A widely accepted definition is that of Timmers (1997 p. 31), "the organization (or 'architecture') of product, service and information flows and the sources of revenues and benefits for suppliers and customers". As such a business model has limited scope and does not include, for example, the overall marketing strategy or general strategic orientation of the firm while it may be concerned with inter-organizational relations. The business plan may later be related to marketing or strategic plans in the implementation phase (Weill & Vitale, 2001). Thus, a business model can be seen as the initial plan that sets the service implementation process on a certain path, which can have implications for the eventual success of the service[1].

Mobile commerce business models, similar to those for e-commerce, will leverage the advantages of a new distribution, sales, and service channel and indeed there is much to learn from the valuable experience e-commerce presents.[2] However, there are aspects of mobile Internet use that make mobile commerce unique. Distinctive characteristics such as ubiquity, accessibility, reachability, localization and personalization create new bases for value (Baldi & Thaung, 2002). Furthermore, these characteristics lead to different settings for value creation: time-critical arrangements, spontaneous decision needs, entertainment needs, efficiency ambitions and mobile situations (Anckar & D'Incau, 2002). Also, the relationship between the end-user and the network operator makes billing and payment functions more convenient.

Leveraging these new sources of value and functionalities leads to a greater emphasis on personalization, and subsequently to more user-centric (Ropers,

2001), or individual or I-centric services (Ballon & Arbanowski, 2002). Furthermore, services are expected to be both passive, where the transfer of data occurs without action on the part of the end user (such as email receipt, status monitoring, and automatic updates) and active, such as shopping, information gathering, and appliance management, which require the participation of the user (Senn, 2000).

In addition to differences in functionalities, mobile commerce will also involve different groups of players than were found in e-commerce. According to functional categories players in the mobile commerce industry include: technology platform vendors, infrastructure and equipment vendors, application platform vendors, application developers, content providers, content aggregators, mobile portal providers and mobile service providers (Tsalgatidou & Pitoura, 2001; Maitland, Bauer, & Westerveld, 2002). One of the main differences in the types of players to date is the absence of Internet Service Providers (ISPs). Although ISPs did not factor directly into the business models of many e-commerce services, they were active in many of the basic elements such as website hosting and providing e-mail services. In mobile commerce these functions have fallen into the domain of the network operator. Another difference between e- and m-commerce is the centrality of the role of middleware providers (Varshney, Vetter, & Kalakota, 2000). Due to the variety of platforms (cHTML, WML, xHTML) and mobile terminals (phones, PDAs, pagers) the effort needed to achieve interoperability is expected to be greater.

2.2 Network Formation in Mobile Commerce

Mobile commerce network formation will occur amidst an economy-wide trend toward production through complex networks. As noted by Hage and Alter (1997, p. 108) "the growth in knowledge and the speed with which it changes has forced organizations toward more complex modes of coordination, greater differentiation of partners, and increased involvement in multiple interorganizational networks." This trend is exemplified in the development of General Motor's OnStar system, a mobile information service, as described by Barabba et al. (2002). In the case the authors describe the complex coordination required for service development and provision, which in its final form involved firms both from inside and outside the industry including auto manufacturers, a media (radio) firm, and content and information service providers. Furthermore, in developing the business model the GM team considered the possible strategic alternatives for prospective competitors and the pros and cons of forming alliances with each group.

As the OnStar example suggests, network formation is a complex process and membership in a network may be based on several criteria. Studies of network formation have concluded that a firm's contribution to end-customer value is of paramount importance in network formation (see Kothandaraman & Wilson 2001; Christensen & Rosenbloom, 1995) and this is likely to be true in mobile commerce as well. Certainly a mobile operator interested in partnering with a bank to develop mobile payment applications will evaluate potential partners based on the value-added they can contribute. It is also likely however that many banks could fill this role and it is often connections, whether social or professional, which differentiate the successful partner. Indeed, these connections have beneficial effects such as discouraging malfeasance and facilitating trust (Granovetter, 1985) as well as contributing to the success of firms (Uzzi, 1996). In mobile commerce where these partners are more likely to come from diverse industries the normal synergies gained from these social connections may be reduced due to inter-industry frictions in cultures and processes.

Thus, the trend toward greater network formation in the provision of services is likely to be observed within the unique circumstances required for mobile commerce. Considering the possible effects of this combination for business models raises the following questions. First, how will mobile commerce networks form and what will be the implications of the distribution of power among the players? Second, how do these networks contribute to the dynamic nature of mobile commerce business models?

3 Relationship-Based Perspectives

Frequently, discussions of mobile commerce business models portray firms as autonomous entities, free to decide which business model to pursue independent of their relations with other firms. In what follows this discourse is expanded by explicitly examining the effects of networks on mobile commerce business models. In particular an examination is made of the distribution of power among firms in a network and how this affects the business model. In a network power is often derived from the contribution to value-added, social or professional contacts or through assets, and power derived in previous business ventures. The discussion of networks, power and mobile commerce business models will take three forms. First, to highlight the role of power and its implications for business models two perspectives or paradigms are proposed. In each paradigm an extreme distribution of power in the mobile commerce industry is represented. Next, the concept of coupling (or joint investments) and its role in terms of power and business models is presented. Finally, the issue of dynamic forces in the power distribution and their affect on business models is considered. After these various perspectives are presented, the concepts are demonstrated through an example of an early mobile commerce venture involving Vodafone and Vizzavi. The section ends with a discussion of the role of power, as presented in the three perspectives, in the development of mobile commerce business models and their implications for content providers.

3.1 Perspectives

In mobile commerce the lack of clarity over the issue of where the value, and hence the power, lies is demonstrated by the uncertainty, at least in some sectors of the industry, over who should pay whom. To better understand the implications of power and explore situations where the source of value is clear, two extreme perspectives are proposed: the 'network perspective' and the 'content perspective'. This is similar to the business model typology consisting of firm-based control, hybrid, and network-based control used by McKelvey (2001) to analyze the relationship between software development processes and broader forms of economic dynamics.

In the mobile commerce context the 'network perspective' represents a market where the value and hence the power lies with those firms controlling access to the network, namely equipment vendors, network operators, or license holders and ISPs. Requesting access are the content providers, application service providers (ASPs) and MVNOs. In this perspective, mobile operators rely on power developed from historical control over access to customers as well as the assets their networks represent. The power of a network operator with their own inter-organizational network will also be influenced by the power it has vis-à-vis other network operators in competing networks of firms.

From the 'network perspective' the formation of inter-organizational networks will be largely at the discretion of those controlling access to the network. In return for access content providers and others may be required to pay fees or agree to exclusive contracts whereby they are restricted from offering their content on other networks. Network operators are allowed to control network access because the mobile marketplace is often a competitive one and thus there is little basis on which policy makers can require open access.

The 'content perspective' or 'content paradigm' represents a market where content is the most highly valued asset in mobile commerce and content firms are able to translate this value into power through which they can direct their peers. In this paradigm operators and Mobile Virtual Network Operators (MVNOs) compete to attract these players to their networks (De Vlaam & Maitland, 2003).

Dominant firms in the content paradigm derive power from the value they bring to mobile commerce but also from the depth of their investments in a variety of media and their brand. Further power derives from the flexibility with which they can enter mobile commerce, as well as leave if necessary. From the content perspective exclusivity means that content from a particular provider will be the only content of that genre carried by a network.

Although these two 'perspectives' or paradigms are merely exaggerations and do not represent realistic mobile commerce scenarios, they provide a framework for analyzing sources and consequences of power. In inter-organizational relations the power will also have an influence on the degree to which firms 'couple' or pursue joint investments. Although coupling and exclusivity are related, for example tight coupling often accompanies exclusive agreements, they are different in that it is possible for firms to be loosely coupled while pursuing exclusive agreements.

Tight and loose coupling are important for business models for several reasons. First, although tight coupling does not wholly determine stability, firms that invest jointly will have a greater incentive to stay together during times of economic or organizational stress. Furthermore, coupling arrangements will have implications for others in the network. Two tightly coupled firms are likely to wield greater power within a network and steer the development of the business model to suit their needs.

The diversity of mobile commerce inter-organizational networks is likely to create instability in the distribution of power and hence it is likely to change. Although within an industry power amongst players usually changes slowly, in diverse networks this is not necessarily the case as the bases for any initial power are less established. As mobile commerce diffuses and sources of value become more well defined (e.g., ringtones), bases of power are likely to change. In turn, changes in the business model may be desired or required.

3.2 Examples

In the following paragraphs an example of an early mobile commerce service provider is presented. The case explores the role of the operator and a content provider in the mobile commerce industry, examines their revenue and business models and notes how the service fits with other firm investments. Subsequently the information will serve as the basis for discussion concerning the role of power in this mobile venture.

3.2.1 Vodafone Group

Vodafone emerged as an independent mobile telecommunications operator in the UK in 1991 and has not looked back. Building on its growth in the domestic arena, in 1993 it began an internationalization drive, joining several consortia and forming Vodafone Group International. By 1999 the company had interests in mobile concerns in over 24 countries and secured a position in the U.S. market through its merger with AirTouch Communications. This position was strengthened in 2000 when the Vodafone AirTouch Group joined forces with Bell Atlantic to launch Verizon Wireless of which the Group owns a 45% stake. Also in 2000, Vodafone Group finalized its purchase of Mannesmann AG, a German telecommunications and engineering conglomerate. The acquisition created an uproar in Europe as it was the first foreign hostile takeover of a German firm. With the acquisition of Mannesmann Vodafone Group nearly doubled in size. In 2001 Vodafone cemented its position in Asia by acquiring majority stakes in both J-Phone and its parent Japan Telecom. Since then Vodafone's international presence has expanded to include 38 countries.

As a conduit for media content Vodafone's strategy, much along the lines of its marketing strategy, is centralized. Vodafone's subsidiaries will all eventually adopt the Vodafone name and are likely to receive a majority of their content through the parent company. Through its Via Vodafone program the company is offering developers access to a gateway that will be reachable by all of their local subsidiaries in return for an unspecified revenue sharing arrangement. These developers will have to compete with other direct Vodafone content investments, such as its purchase of football media rights from KirchMedia and a marketing (and content development) agreement with David Beckham of Manchester United. This is in addition to the legacy of the Vizzavi organization, Vodafone's content portal.

3.2.2 Vizzavi

Vizzavi was a portal formed in 2000 through a $1.4 billion, 50/50 joint venture of Vodafone Group Plc. and Vivendi Universal. Vodafone is one of the largest mobile phone companies worldwide and Vivendi Universal is a global media and communications firm, which among other activities owns Universal Music Group as well as Canal+.

In creating Vizzavi, the partners aimed to develop a multi-access portal, to be used by fixed and mobile customers alike. In terms of mobile commerce, Vizzavi served as a branded content aggregator/portal and was to be the initial default portal for subsidiary operators of both Vodafone and Vivendi. The formation of Vizzavi caused apprehension at the European Union where policy makers were concerned that rival portals might be excluded from Vivendi's set-top boxes and Vodafone's mobile terminals. Approval was eventually granted based on the special condition that the consumer be able to change the default portal to one of his or her own choosing (CIT, 2001).

As a content aggregator, Vizzavi made agreements with a variety of firms, including Google, Reuters, and eCentive. Despite these agreements the venture has been a disappointment to both parent companies, having failed to win enough subscribers and generate substantial revenues. Existing web portals and mobile multimedia GPRS services apparently created more competition than the service could handle. Vizzavi intended to meet this challenge by focusing greater attention on the acquisition of visual content.

The ownership structures that governed the relations of Vizzavi with its clients are complex. As the default portal for Vodafone's and Vivendi's networks throughout Europe, it was meant to serve as a source of content for these operators. The names, locations, and Vodafone's stake in selected operators during 2001 are shown in Table 1. In each operating territory a local Vizzavi entity was established to provide language-specific and culturally relevant content. This local company was owned 80% by Vizzavi Europe Ltd. and 20% by the local network operator. These relations are depicted in Figure 1.

Table 1: Vizzavi Portal Presence[3]

Operating country	Year of Agreement	Local network operator	Vodafone Stake
UK	2000	Vodafone	100%
Netherlands	2000	Libertel/Vodafone	70%
Germany	2001	Vodafone Mannesmann	99%
Italy	2001	Omnitel Vodafone	76.1%
Greece	2001	Panafone	55%
Portugal	2001	Telecel/Vodafone	50.9%

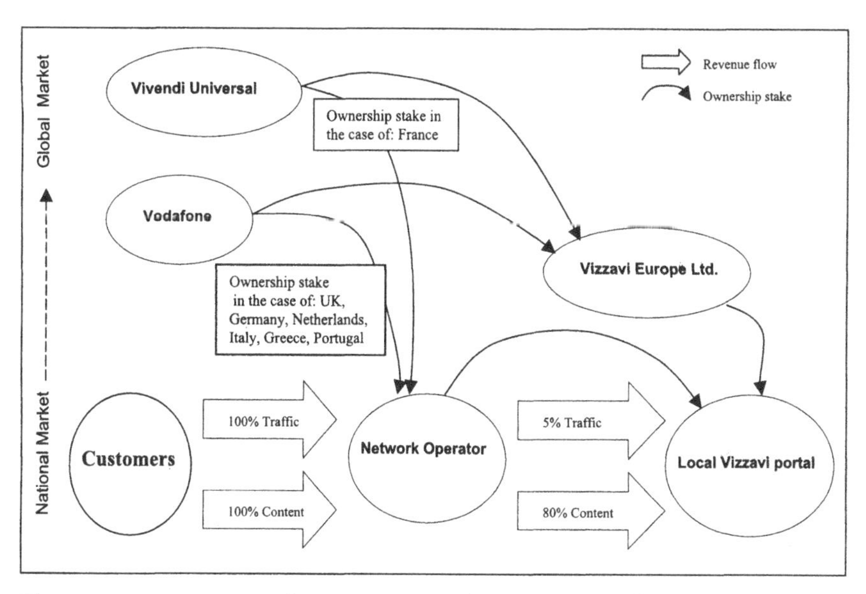

Figure 1: Vizzavi Ownership Structure and Revenue Model
(prior to August 2002 Vivendi divestiture in Vizzavi)

The complexity of the ownership structure is replicated in the revenue model, which has two components. First, operators and Vizzavi split the portal-generated traffic revenue 95/5. Traffic is generated by accessing free information located at the portal or through the use of premium content such as ringtones, logos, games and downloads. The premium content was available only to those who subscribed to the individual services. The revenues from the end users for these services were to be split 20/80 between operators and Vizzavi.

The revenue model in the original agreement stipulated that the portal and mobile operators were to split the gross margin 50/50. A revised model was explained as an adjustment to reflect a greater emphasis on premium content as both a source of revenue and an integral part of mobile services. Although the change guaranteed Vizzavi a greater share of the revenues from premium content subscriptions, it lost a share of the traffic revenue presumably driven by the quality of its content.

The ruling by the EU canceled any hopes that Vizzavi may have had for being the exclusive portal for Vodafone and Vivendi's mobile outlets. Forced to compete with other portals Vizzavi gave exclusive deals to some content providers (i.e., Google) and obtained exclusive deals from others (i.e., Universal's artist Sting).

In the end, despite changes in the revenue model and cost cutting measures to improve its performance, the venture did not last. In August 2002, Vivendi under pressure to reduce its debt sold its 50% stake in Vizzavi to Vodafone. Vodafone subsequently integrated Vizzavi into its group and has relabeled it Global Content Services.

3.3 Discussion

In this example one sees that power between network operators and content providers shifts in an attempt to strike a balance between the extremes of the network and content paradigms. In the Vizzavi example, revenue is being shared with the content providers as opposed to the early stages of WAP where network operators required content providers pay for access to their networks (Kar, 2002). Since Telia began offering revenue sharing for SMS service providers and the i-mode model was successfully implemented, revenue sharing has gained popularity and power is shifting toward the content providers. Despite this trend, the behavior of Vodafone and commentaries by Geng and Whinston (2001) and Funk (2002) suggest that this transfer of power is at the discretion of network operators.

The example also represents a case of tight coupling and its implications for business models. On the one hand the tight coupling through an elaborate ownership structure resulted in a somewhat elaborate business model for Vizzavi. As a mobile portal the tight coupling with Vivendi provided them with access to their parent company's content, although not uniformly on an exclusive basis. As with any joint venture, Vizzavi can be either the beneficiary or the victim of its parents' status. In this case the tight coupling did not produce the anticipated advantages which eventually led to Vizzavi's demise.

This interdependency to quickly expand with little concern for the relationships between operators. Although Vizzavi, a portal, and content providers are not directly comparable, one can conclude that the sharing of traffic revenues for Vizzavi was possible due to their relationships with the operators, while it is unlikely that content providers will manage traffic-dependent revenue agreements.

The last issue considered is dynamics in power. As the move away from the network paradigm that existed with WAP service suggests, power is shifting toward content providers, expanding their options. However, as is demonstrated by the case of Vizzavi, power within a network can also change and require revision of the business model. The power defined by the added end-customer-value can be eroded by exogenous factors like large debt that can swing the balance of power back toward the network operator. Thus, it is likely that over time, as various players gain experience, develop value, or gain a better understanding of their options that models developed within a network will change.

Finally, the consideration of the balance of power, tight and loose coupling, and dynamics in power will all have implications for content providers. While some operators such as Vodafone have taken a centralized approach to content approval and distribution other operators, such as those using NTT DoCoMo's i-mode model, have opted for a decentralized approach. Deciding which scenario is beneficial for the content provider will require an understanding of the potential level of coupling between the two parties, the power they have vis-à-vis a particular operator, which will likely be based on the value they provide to the overall mobile data service, their personal connections, and finally an informed view of what the changes in the balance of power are likely to bring.

4 Conclusion

The development of mobile commerce business models, although informed from experiences with e-commerce, takes place in a unique environment with new sources of user-value delivered through services that are offered through networks of diverse firms. Due to the diversity of these networks the distribution of power is likely to be unstable. The implications are diversity and instability in the business models. Changes in business models will be observed on an industry-wide basis, as portals, content providers and middleware developers quickly expand across markets around the globe. Changes will also be seen within networks as the roles and relations among partners change.

This fluidity in business models can have positive effects for the industry as long as end-users remain unaffected. The possibility to gain power and attain a more favorable negotiating position may provide a strong incentive for innovation. Since the mobile commerce industry is young and has short time-to-market cycles, these processes appear to be complementary.

Although business models are derived from a wide variety of influences, the focus here has been on inter-organizational networks and their accompanying power distributions. Other plausible explanatory factors that may explain the behaviors observed here include changing asset and stock valuations, technological trends, and broader corporate strategies. Thus there is a need for future research that explicitly compares these factors to allow for better understanding of the relationship between the mechanisms of inter-organizational networks and the business models through which they offer products and services.

Endnotes

[1] For an example in the e-commerce realm see Gallaugher et al. (2001).
[2] Compare, for example, the mobile commerce business models proposed by Tsalgatidou and Pitoura (2001); (content providers, mobile portals, gateway providers, service providers) with the atomic e-commerce business models identified by Weill and Vitale (2001). Also note, for example, the warning by Anckar and D'Incau (2002) that firms involved with mobile commerce do not assume that high levels of mobile phone penetration signal a high level of acceptance for m-commerce, as some e-commerce actors assumed high levels of Internet and personal computer (PC) translated to an acceptance of electronic commerce.
[3] Source: CTI 2001 and various corporate websites.

References

Anckar, B., & D'Incau, D. (2002). Value-added services in mobile commerce: An analytical framework and empirical findings from a national consumer survey. *Proceedings of the 35th Hawaii International Conference on System Science-2002.*

Baldi S., & Thaung, H. P.-P. (2002). The entertaining way to m-commerce: Japan's approach to the Mobile Internet—A model for Europe? *Electronic Markets, 12*(1), 6-13.

Ballon, P., & Arbanowski, S. (2002). Business models in the future wireless world. WWRF WG2 White Paper. *World Wireless Research Forum.* www.wireless-world-research.org/.

Barraba, V., Huber, C., Cooke, F., Pudar, N., Smith, J., & Paich, M. (2002). A multi-method approach for creating new business models: The General Motors OnStar Project. *Interfaces, 32*(1), 20-34.

Christensen, C., & Rosenbloom, R. S. (1995). "Explaining the attacker's advantage: Technological paradigms, organizational dynamics, and the value network." *Research Policy, 24,* 233-257.

CIT (2001). *3G in Europe: Future Markets (Preparing for launch).* CIT Publications Ltd.: Exeter, UK.

Dehghan, S., Lister, D., et al. (2000). W-CDMA capacity and planning issues. *Electronics & Communications Engineering Journal, June,* 101-118.

De Vlaam, H. & Maitland, C. (2003). Competitive mobile access in Europe: Comparing market and policy perspectives. Retrieved from the World Wide Web: http://faculty.ist.psu.edu/maitland/deVlaam_Maitland.pdf.

Funk, J. L. (2002). Japanese Mobile Internet Navigation Services: Competition between train information, map, and destination information providers. *Working paper, Research Institute for Economics and Business Administration,* Kobe University, Kobe, Japan.

Gallaugher, J. M., Auger, P., & BarNir, A. (2001). Revenue streams and digital content providers: an empirical investigation. *Information & Management, 38,* 473-485.

Geng, Z., & Whinston, A. B. (2001). Profiting from value-added wireless services. *IEEE Computer, 34*(8), 85-87.

Granovetter, M. (1985). Economic action and social structure: The problem of embeddedness. *AJS, 91*(3), 481-510.

Hage, J. & Alter, C. (1997). A typology of interorganizational relationships and networks. In J.R. Hollingsworth & R. Boyer (Eds.), *Contemporary Capitalism: The embeddedness of institutions* (pp. 94-126). Cambridge: Cambridge University Press.

Kar, v.d. E. A. M. (2002). The Introduction of M-Info—a teaching case. *Proceedings of the 15th Bled Electronic Commerce Conference. Bled, Slovenia,* June 17-19, 650-671.

Kothandaraman, P. & Wilson, D. T. (2001). The future of competition: value-creating networks. *Industrial Marketing and Management, 30,* 379-389.

Maitland, C., Bauer, J. & Westerveld, R. (2002). The European Market for Mobile Data: Evolving Value Chains and Industry Structures. *Telecommunications Policy, 26,* 485-504.

McKelvey, M. (2001). The economic dynamics of software: Three competing business models exemplified through Microsoft, Netscape and Linux. *Economics of Innovation and New Technology, 10*(2-3), 199-236.

Ropers, S. (2001). New business models for the mobile revolution. *EAI Journal, February,* 53-57.

Senn, J. A. (2000). The emergence of m-commerce. *Computer, December,* 148-150.

Timmers, P. (1999). *Electronic Commerce: Strategies and models for business to business trading.* Chichester UK: John Wiley and Sons.

Tsalgatidou, A. & Pitoura, E. (2001). Business models and transactions in mobile electronic commerce: requirements and properties. *Computer Networks, 37,* 221-236.

Uzzi, B. (1996). The sources and consequences of embeddedness for the economic performance of organizations: The network effect. *American Sociological Review, 61*(August), 674-698.

Varshney, U., Vetter, R. J., & Kalakota, R. (2000). Mobile Commerce: A new frontier. *Computer, October,* 32-38.

Weill, P. & Vitale, M.R. (2001). *Place to Space: Migrating to e-business models.* Boston: Harvard Business School Press.

12 Mobile Communications Business Model in the United States

James Alleman & Christopher Swann

1 Overview

The worldwide mobile communications industry is at a crossroads. The voice market has been rapidly maturing. The 3G technology is unproven, and has experienced technical difficulties and delays; its cost is higher than the previous 2G system, and moreover, spectrum is insufficient and expensive. This chapter will review the issues facing the industry and their implications for a mass media content strategy. We discuss the United States, including its particularities, but the main dynamics apply to other developed countries, too. The chapter is divided into five sections. This section provides a brief overview of the industry. The next two sections discuss the major problems and issues confronting the U.S. cellular industry: spectrum cost and allocations, standards, and pricing strategies for voice services.[1] The fourth section summarizes the business model practiced by the wireless carriers. The last section reviews the industry and concludes.

1.1 Growth/Revenue

The growth of the U.S. wireless telecommunications industry has been remarkable, similarly to other countries with average rates of more than 25% per annum since 1993. As seen in Figure 1, the number of subscribers is more than 170 million, and has reached more than 60% of U.S. households.[2] Although revenues have grown with this increase in penetration to more than $90 billion, until recently the average revenue per customer (ARPU) has actually declined. The proliferation of service providers undermined the high prices and limited service of the previous duopoly and increased price competition.[3] The recent ARPU increase appears to be the result of data service revenues (Standard & Poor's, 2004, p. 8). Roaming revenues—the subscriber's

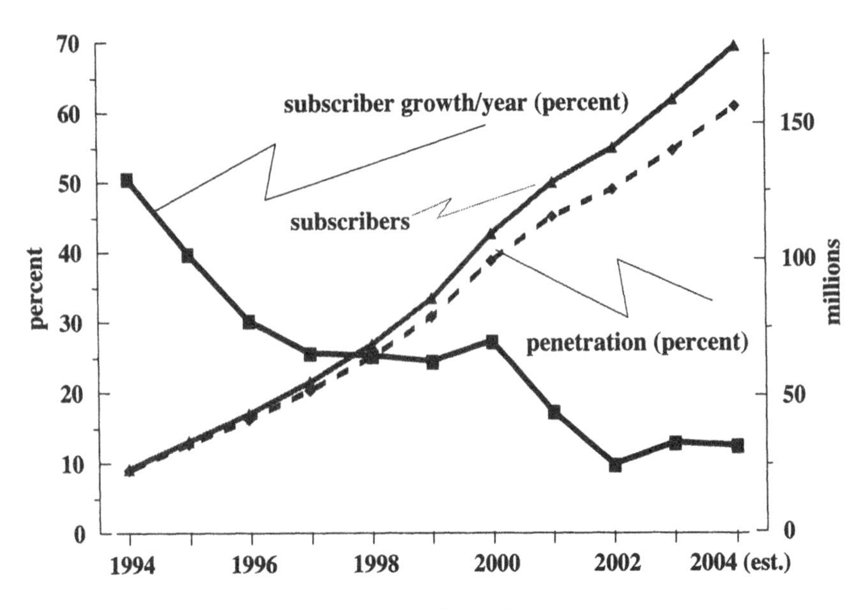

Figure 1: Subscribers, subscriber growth, and penetration
Source: [S&P, 2001 & 2002]

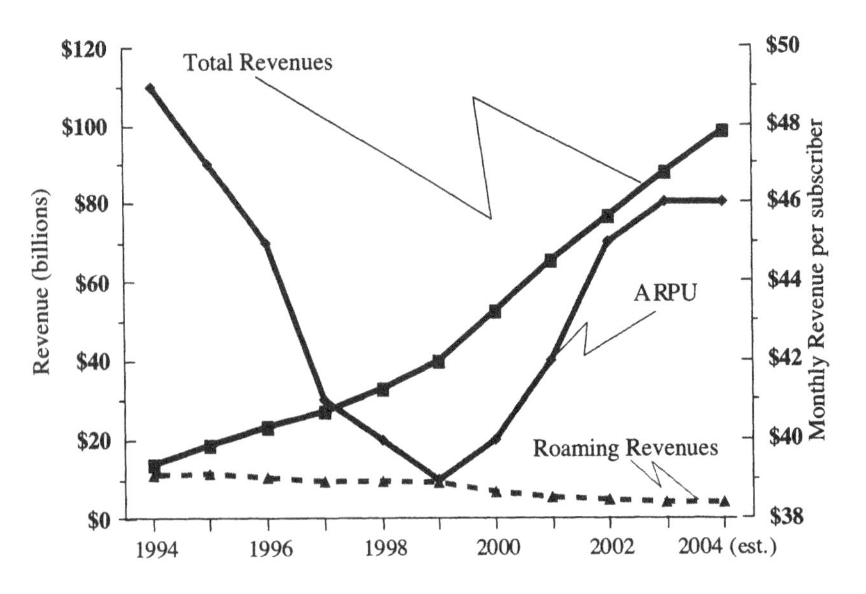

Figure 2: Total revenue, average revenue per subscribers, and roaming revenu
Source: [S&P, 2001 & 2002]

charge for calls initiated or received from outside the designated service area —declined (see Figure 2), and are being eliminated as a competitive response. With lower revenues per customer it becomes harder to subsidize handsets. These issues will be addressed below.

1.2 Market Players (USA)[4]

The three major wireless voice providers: Cingular (including AT&T Wireless since 2004),Verizon Wireless, and Sprint (including Nextel since 2005), control nearly 80% of the market; Alltel (which acquired Western Wireless in 2005), US Cellular and T-Mobile USA hold the balance (Belson, 2004, 2005). Figure 3 summarizes the carriers' market share.

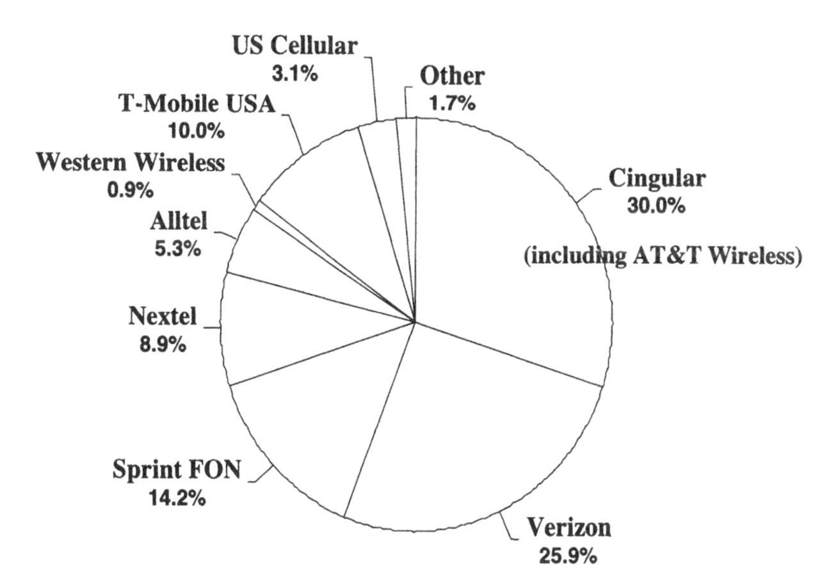

Figure 3: United States Wireless Market Share (Q2 2004)
Source: Standard & Poor's Industrial Survey: Telecommunications Wireless November 4, 2004, p. 8

2 What are the Issues?

Spectrum allocations, geography, technologies, and price policies are issues that distinguish the United States from the European environment. We will address each of these in turn.

2.1 Spectrum Issues[5]

Inadequacy

The United States wireless industry believes that the allocation of electromagnetic spectrum is inadequate for its needs. Whereas Europe has allocated 155 MHZ of spectrum for 3G components (NTIA, 2002), the U.S. has only allocated 90MHz (FCC, 2002b). There are plans to auction more spectrum from the UHF TV band, but the television industry has resisted.[6]

Regulatory Delays

United States wireless had additional barriers. The pre-divestiture Bell System experienced significant regulatory delays which allowed the rest of the world to jump ahead on mobile communications (Hausman, 2001; Noam, 2001). All this cost the United States' consumer an estimated $20 billion in welfare loss (Crandall, 2000). The wireless service went to the regional Bell Operating Companies.[7] In the FCC's desire to perfect the auction system for second generation mobile, it delayed the introduction of digital service, allowing Japan and Europe to move ahead (Noam, 2002).

Thus, the US has had a lower penetration of wireless than some other counties. In addition to the delay, the US has a geographic vastness that other countries do not have, inexpensive landlines, and multiple standards.

2.2 Multiple standards[8]

Europe, and much of the rest of the world, adapted the Global System Mobile Communications (GSM) standard, which allows for inter-country roaming and economies of scale in the production of handsets; in contrast, multiple standards exist for mobile service in the United States. This locks the subscriber into the service provider, since the handset cannot be used with a new provider who uses a different standard.[9] Because the handsets are heavily subsidized, lock-in has benefited the wireless providers, although it is a loss to customers. Customers also must agree to a one-year contract when they initiate service. In addition to locking-in the customer, this ensures that at least a portion of the handset subsidy is covered by customers' monthly recurring charges.

Multiple standards have the advantages of competition among technologies, and barriers to some oligopolistic consolidations. If companies wish to consolidate their networks or share facilities, it is more difficult to do so with different standards. It also limits the ability of any US company to be a global player on two levels. First, it would need different technical skills and capabilities if it wished to enter different markets outside of the US. Second, it limits its international roaming revenues.[10]

3 Pricing Strategies

In 1998 AT&T Wireless introduced its "one-rate" calling plan. It changed the pricing dynamics of the industry. The one-rate plan allowed AT&T wireless subscribers to use their minutes anywhere in the United State—without roaming charges or distance charges. Since AT&T's footprint was nationwide, it represented a significant threat to the regional carriers. This led to consolidations, sharing and joint ventures to develop nationwide coverage and combat loss of market share to AT&T. The carriers began to offer plans with a large number of "free" minutes. Moreover, the average price per minute continues its historical decline—13% between 2000 and 2001 and 9% between 2001 and 2002 (Standard & Poor's, 2004, p. 9). The pricing plans are similar. The game is to get the customers, and to lock them in with a contract and a handset that cannot be used with other carriers.

When one examines the tariffs of the major carriers, then, price differences do not appear all that significant. However, the myriad of options that are offered to customers confuses them. Massive amounts of "free" weekend minutes and "free" minute allowances with different monthly charges make the plans difficult to compare with one another. It allows the carriers to compete on a basis other than price. Brand name and reputation play a predominant role.

Moreover, the choice of a provider is not easy for the consumers. With the different pricing options, the consumer is forced to estimate minutes, something that is difficult to do (Alleman, 1984; Paverini, 1979). If consumers had perfect information, these self-selecting rates could be welfare enhancing (Willig,1978); however, this is not the case. On average, customers over-estimate their usage; thus, they pay more than they would if they had correctly estimated their minutes. It is particularly hard to estimate the impact of the tariff under receiving party pays, since, in addition to estimating her own traffic, the customer has to estimate the number of incoming calls. Thus, these self-selecting tariffs are revenue enhancing for the carriers.

3.1 Pricing and Marketing War

What is the incentive for the carriers to price so aggressively? First, the cost structure supports it. The industry has a high fixed, and mostly sunk cost with much lower variable costs. This means that they can drive their cost close to the incremental cost, because it will at least cover some of the fixed cost. Once the carrier captures a customer there is less need to price so low. But, in order to acquire the customer a carrier needs to offer enticements. If a carrier fails to capture the potential customer, it will be difficult to entice her away from a competitor because of the lock-in. As a result of price competition, wireless prices, per minute of usage, are much lower than in Europe and Japan, and usage minutes per subscriber much higher. But is this pricing strategy sustainable? Two factors are undercut it. First, number portability became available in 2003. Second, the move to 3G will only have two standards, at most: WCDMA and CDMA2000. Thus, the ability to maintain lock-in with the handset should be diminished.[11]

4 Is there any hope? Data? 3G?

Data

The main hope for the carriers is a boom in data services. European providers have had success with short messaging services. DoCoMo, in Japan, has had even greater success with its i-Mode Internet data service. I-Mode is a "2.5" G service. It features an always-on connection much like a cable modem or DSL connection in the wired world. Also, the service is priced by the bits transferred, not air minutes. This can make a significant difference in the total cost to the consumer. As a result, it has captured 60% of the market based on this technology and is the second largest Internet Service Provider in the world. It should be noted that the demand for 3G in Japan has been below expectations (Belson, 2002). This does not bode well for the US carriers, or the European carriers for that matter.

In the United States the adoption of data services delivered over the mobile phone has been slower. With the exception of teenagers, consumers have not readily accepted the early data services offered on their handset. The technology is still awkward to use, and close substitutes are readily available. Moreover, in the carriers' closed systems, they control the available content. Thus, it has only been a small percentage of the carriers' revenue. In 2000 it was $211 million—negligible; in 2001 it was $545 million less than one percent of total revenue. However, the market has grown much larger with new offerings such as ringtone downloads, picture-mail, and particularly text messaging or Short Messaging Service (SMS). Data revenues for 2004 are estimated at $4.3 billion (Yankee Group, 2004; Global Insight, 2005). Nevertheless, the level of revenues from these services remains a small portion—on the order of 4%—of total carrier revenues.

WiFi, a wireless service targeted mainly to laptop computers, has spread in the urban areas. This allows people to sit in coffee shops, airport lounges, or other "hot spots" to access their e-mail and web services at high-speed data rates. This service is competitive to wireless data services on mobile phones.[12] Blackberry, a wireless service that allows users to send and receive e-mail while on the go, is another competitor to voice-mobile data services with 800,000 dedicated higher-value customers. T-Mobile has introduced a service in 2003 which combines BlackBerry and voice features to mitigate this threat. In addition it has pursued a strategy to sell WiFi service in Starbucks and other locations as part of its service offering. Other mobile providers are expected to follow a similar path (Standard & Poor's, 2004, p. 14). Given the current lack of demand for data, close substitutes, and no "killer" application on the horizon, it is problematic whether the wireless voice industry can be saved with data services.

Next Generation 3G

The industry, worldwide, has touted 3G as the great leap forward (and portends a similar result to its Chinese namesake). It has better quality of service; it can handle data better—via packet switching. It handles spectrum more efficiently. However, at least two major problems are associated with 3G. The first is the transition of current subscribers from 2G to 3G. The handset, spectrum, and system are all more expensive than the current system. This means that the subscribers' cost will be higher. New mobile customers are unlikely; most probably the customers will come from their existing subscriber base. What can carriers do to migrate their customers to the next generation, if they are happy with their service, and are not enthusiastic to pay more for a service, which does not offer much perceived improvement? Not much, it would seem. But, for argument's sake, assume that the carriers are successful in migrating their customers. Then the existing 2G-business collapses. It is a lose–lose situation. If the firm successfully migrates subscribers, it loses the revenue from the older service without an offsetting cost savings. On the other hand, if they fail to move customers from 2G, their investment in the 3G service is lost.

But in a market where all carriers will be going after the same customers, a price and marketing war is likely to continue with even greater intensity than before.

5 Outlook and Conclusions

While the indications are that mobile cellular is a maturing industry, much uncertainty remains. In the past the wireless carriers have been able to implement a lock-in strategy. To lock the subscriber in the carrier must first obtain the customer. The marginal cost of service for each additional customer is low, particularly compared with the large sunk cost of the network, but the acquisition cost is high, thus the drive to obtain the customer is intense. And when 3G services begin, the drive to capture customers will become ever more intense. Carriers will attempt to poach each other's markets. Price wars and intense marketing campaigns will ensue. With the introduction of 3G, it is an industry in which competition has worked to the benefit of the consumer, but to the detriment of the carriers.

This will lead to bankruptcies and consolidation of the industry. Indeed, consolidation may be one of the only strategies that the wireless carriers can implement and win by reducing price competition. The mergers of Cingular/ AT&T, Sprint/Nextel, and Alltel/Western Wireless, all in rapid succession, are past of these scenarios of consolidation. Given that the top three firms now account for 80% of the market, the consolidation strategy is reaching its limits. The second major strategy for wireless carriers is to increase their for individualized and mass media type content. With voice minutes and subscriptions approaching saturation, this seems to be the main avenue of growth, in America and worldwide.

Endnotes

[1] Unless otherwise stated, all of the references are to the United States voice-mobile market.

[2] Seventy percent penetration has been estimated as the point at which growth will be virtually satiated, although this may be high for the United States (Shere, 2001).

[3] The Federal Communications Commission (FCC), at the time, seemed to think that a duopoly would have impact associated with the economic concept of competition, that is, prices at or near marginal cost; no monopoly rents, *et cetera*. This was far from the case. The regulatory authorities in the United Kingdom and elsewhere made similar errors. See Swann (2002).

[4] As indicated earlier, we will not address the wireless data market except as it is integral to the voice providers' offering or if it is in competition to wireless voice.

[5] Although not addressed here, auctions have a perverse effect on the allocations. The rationale for auctions is to allocate the resource to its best use. However, government, at least in the recent past, has assumed that proceeds of the auctions accrue to the government. This has led to inefficient behavior in setting up the auction and does not account for the adverse tax effects (Noam, 2002; Alleman, 2002).

[6] For a detailed discussion of the 3G spectrum issues, see NTIA (2002).

[7] AT&T was not reluctant to give this up, since an internal report done for it by McKinsey in 1981 indicated that the demand for mobile service would not exceed 900,000 by 2001, far below the current 170 million mobile subscribers in the United States today.

[8] For a brief history of wireless development, see Liew (1999).

[9] It is not simply that the standard locks in the customer. In the US the service provider has a veto over the handset used. It adds its own requirements to the handset used, in contrast to Europe where the handsets are interchangeable among carriers. Noam (2002) noted that this type of vertical integration is greater than in any other industry. As he puts it "The subscriber is 'owned' by the carrier." In the data arena, the carriers control the content available to customers.

[10] T-Mobile was the only exception; it provides a dual-mode handset that is capable of operating in countries that have the GSM standard. Subsequently, Cingular (including AT&T Wireless) offer GSM platforms (Standard & Poor's, 2004).

[11] Dominant wireless providers can impose constraints and codes in the handset to make them incompatible with similar technologies, but regulatory control over these actions can eliminate these anti-competitive acts. See Noam (2002) for an analysis of vertical control in the industry and means of alleviating it.

[12] Antenna technology has been developed to extend the range of 802.11b signals to up to four miles. If this is successful, it has serious negative implications of the data space for the wireless voice service providers.

References

Alleman, J. (2002). *A Note on the Inefficiency of Spectrum Auctions.* Working paper (available from the author).

Alleman, J. & Schmidt, L. W. (1984). Telecommunications in a Fickle Regulatory Environment. In P. C. Mann & H. M. Trebing (Eds.), *Changing Patterns in Regulation, Markets, and Technology: The Effect on Public Utility Pricing,* Michigan State University, East Lansing, Michigan.

Belson, K. (April 22, 2002). Japan Slow To Accept New Phones. *New York Times,* p. C-4.

Belson, K. (December 16, 2004). Latest Merger Would Recast Cellular's Face Once Again. *New York Times,* p. C-1.

Belson, K. (January 11, 2005). Alltel to Buy Western Wireless in $6 Billion Deal. *New York Times,* p. C-1.

Cellular Telecommunications & Internet Association (CTIA). (2002). Retrieved from the World Wide Web: http://www.wow-com.com/.

Crandall, R. W. & Hazlett, T. W. (2000). *Telecommunications Policy Reform in the United States and Canada,* AEI-Brookings Joint Center For Regulatory Studies, December.

De Aenlle, C. (June 2, 2002). Beyond the Big Loss of Vodafone. *The New York Times.*

Federal Communications Commission (FCC 2002a). *Broadband PCS.* May 23.

Federal Communications Commission. (FCC 2002b). *Spectrum Policy Task Force Presents Recommendations for Spectrum Policy Reform.* November 7.

Federal Communications Commission (FCC 2002c). Retrieved November 5 from the World Wide Web:http://wireless.fcc.gov/services/broadbandpcs/operations.

Global Insight. (2005). Unpublished forecast of Telecom/IT Group.

Hausman, J. (2002). Wireless From 2G to 3G: Competition for Internet-Related Services. January 22.

Liew, J, *et al.* (2002). Wireless Voice To Data: The Impact on the Consumer. Retrieved from the World Wide Web: http://www.ksg.harvard.edu/project6/.

Noam, E. (forthcoming). Opening the 'Walled Airwave, in this volume.

Noam, E. (2005). *Ownership and Concentration in the US Communications Industry.* New York: Oxford University Press (forthcoming).

National Telecommunications and Information Administration (NTIA). (2002). An Assessment of the Viability of Accommodation Advance Mobile Wireless (3G) Spectrum in the1710-1770 MHz and the 2110-2170 MHz Bands, July 22. Retrieved from the World Wide Web: http://www.fcc.gov/3G/3Gva072202.pdf

Pavarini, C. (1979). The Effect of Flat-to-Measured Rate Conversions on Local Telephone Usage. In J.T. Wenders (Ed.), *Pricing in Regulated Industries II.* Denver, CO: Mountain States Telephone Co.

Schiesel, S. (January 13, 2005). For Wireless, the Beginnings of a Breakout. *New York Times.*

Shere, C. K. (2002). Telecommunications: Wireline. *Standard & Poor's Industry Surveys,* May 30.

Standard & Poor's. (2004). Standard & Poor's Industry Survey—Telecommunications: Wireless, November 4.

Swann, C. (2002). Presentation at Global Insight's World Outlook Conference, New York, NY, October 30.
Tachkawa, K. (2002). NTT DoCoMo and the New Global Communications Community, Presentation Columbia Business School, New York, NY, November 13.
Willig, R. (1978). Pareto Superior Non-linear Outlay Schedules. *Bell Journal of Economics*, 55-69.
Yankee Group. (2004). Market for wireless data services remains untapped. *Silicon Valley/San Jose Business Journal*, December 27. from the World Wide Web: http://www.bizjournals.com/sanjose/stories/2004/12/27/daily7.html.

13 Mobile Wireless Strategy of Media Firms: Examining the Wireless Diversification Patterns of Leading Global Media Conglomerates

Sylvia M. Chan-Olmsted & Byeng-Hee Chang

1 Introduction

Just as in the oil and automotive industries in the 20th-century, the media industry is going through a profound transformation, moving from a primarily national to a global commercial-media market, and in the process creating a group of global oligopolists (McChesney, 1999). While this trend of global conglomerization continues in the media industry, the new platform for content distribution, mobile wireless, has expanded rapidly around the world, suggesting ample opportunities for these global media companies. In fact, while the demand for traditional media is saturating in many developed countries, wireless penetration was expected to reach 30% of the world's population by 2007, with certain regions such as Asia-Pacific growing at the highest annual rate of 13.6% (Greenspan, 2004). The growth of wireless services in regions such as Western Europe and many of the emerging economies presents an attractive business opportunity for the leading media conglomerates as they attempt to diversify internationally and into other new media businesses. This chapter assesses the product and international diversification strategy of the leading global media conglomerates in the mobile wireless market. Based on the strategic management literature in diversification and a review of the market characteristics and trends in the international media market, we also suggest a system of drivers that influence the conglomerates' diversification strategy into this particular sector.

2 Diversification Literature

Scholars have suggested that the development of global media conglomerates is driven primarily by the privatization of television in many European and Asian markets, deregulation of media ownership, increasing lifestyle parallelism, saturating demands for many media products in the U.S., and the advance of new communications technologies (McChesney, 1999; Chan-Olmsted & Albarran, 1998; Hollifield, 2001; Noam, 2005).

To provide a framework for our analysis of the conglomerates' diversification strategy, we will first review a body of literature that addresses the concepts of diversification, specifically product diversification, geographic market diversification, and the interrelationship between product and geographic diversification.

2.1 Product, Geographic, and Product Geographic Diversification

Diversification has had a rich tradition as a topic of research since the late 1950s (Chandler, 1962; Gort, 1962; Ansoff, 1957, 1958). While Berry (1975) defined "diversification" as the extent to which a firm is active in a number of industries, Booz, Allen, and Hamilton (1985) more specifically referred to "diversification" as a means of spreading the base of a business to achieve improved growth and/or reduce overall risk which may take the form of investments that address new products, services, customer segments, or geographic markets.

Salter and Weinhold (1979) proposed three general but related models in the discussion of corporate diversification strategies. The product/market-portfolio model emphasizes the attractiveness of the target market in terms of attributes such as market size, growth rate, and profitability. The strategy model stresses the interrelation between the core-business market and the target market, which is the emphasis of this chapter. The third approach, risk/return model, derives mainly from financial theories and reflects the concern and interest of investors. Studies of diversification have generally focused on one or more of the three aspects of diversification: (1) the "extent" (i.e., less or more diversification), (2) the "directions" (i.e., related or unrelated diversification), and/or (3) the "mode" (i.e., diversification via internal expansion/mergers and acquisitions or choices of M&A strategy) of diversification (Qian, 1997; Sambharya, 1995; Miller & Shamsie, 1999).

Diversification strategy may be studied either from the "product" or "geographic" perspective. More recent studies in product diversification often investigate the directions of diversification as related or unrelated (Rumelt, 1984; Qian, 1997). Some have argued that related diversification might exploit economies of scope, product knowledge, and other relevant experience, thus reducing transaction costs and improving performance (Williamson, 1981; Grant, 1988). Others have found no differences or the opposite (Grant & Jammine, 1988; Michel & Shaked, 1984). In general, the resource-based view of strategic management strongly argues for strategic relatedness within a conglomerate when it comes to diversification strategy (Chatterjee & Wernerfelt, 1991).

International market or geographic market diversification may be defined as when a firm is horizontally and vertically integrated across different national sub-markets (Hisey & Caves, 1985). The benefits of diversifying internationally originate from two sources—greater opportunities for higher returns and lower correlations of assets across countries (Cavaglia, Melas, & Tsouderos, 2000). Research has shown that international diversification provides firms with significant advantages, including better firm performance (Hitt, Hoskisson, & Ireland, 1994; Tallman & Li, 1996; Grant & Jammine, 1988; Kim, Hwang, & Burgers, 1993). Several studies have suggested that international diversification results in superior performance because it leads to stability of returns, as well as economies of scale, scope, and experience (Caves, 1982; Kogut, 1985; Kobrin, 1991).

As for the interrelationship between international and product diversification, some research has shown that both international and product diversification individually have no effect on firm performance but their interaction leads to a substantial increase in firm performance (Sambharya, 1995). Hitt et al. (1997) found that geographical diversification improves performance in firms that are highly diversified in terms of product markets. In terms of the directions of diversification, some have advocated that relatedness is especially important as the utilization of core skills, know-how, and management resources is necessary in reducing uncertainties in the process of internationalization (Qian, 1997). Nevertheless, studies have also indicated an inverse relationship between product and international diversification (Grant & Jammine, 1988; Buhner, 1987; Madura & Rose, 1987). As both types of diversification involve substantial risks, it's unlikely that a firm would take on both strategies simultaneously. Thus, firms that are diversified internationally would be less diversified in terms of products (Shambharya, 1995). In sum, scholars have consistently concluded that geographic and product diversifications interact with one another and, individually and collectively, influence differential firm performance (Miller & Pras, 1980; Montgomery, 1982; Palepu, 1985; Grant, 1987).

2.2 Product and Geographic Relatedness and Complementary Resource Alignment for Global Media Conglomerates

The type of diversification one would expect to result from a resource depends on its specificity within a particular industry (Chatterjee & Wernerfelt, 1991). The major distinction between media and non-media products rests in the unique combination of the following media characteristics. First, media conglomerates offer dual, complementary media products of "content" and "distribution." Second, media conglomerates rely on dual revenue sources from consumers and advertisers. Third, most media "content" products are non-excludable and non-depletable "public goods" whose consumption by one individual does not interfere with its availability to another but adds to the scale economies in production. Fourth, many media "content" products are marketed under a windowing process in which a "content" such as a theatrical film is delivered to consumers via multiple outlets sequentially in different time periods (e.g., satellite television pay per view, pay cable network, and broadcast network). Finally, media products are highly subjective to the cultural preferences and existing communication infrastructure of each geographic market/country and are often subject to more regulatory control from the host country because of their pervasive impacts on individual societies.

The listed characteristics of media products lead to a market environment in which related product/geographic diversification as well as complementary resource alignment are likely to be the preferred diversification strategy. The symbiotic relation between media content and distribution products presents a classic case of resource alignment. The fact that an existing product may be redistributed to and reused in different outlets via a windowing process reinforces the advantage of diversifying into multiple related distribution sectors in various international markets to increase the revenue potential for such a product. The dual-revenue source mechanism creates another driver for related and complementary diversification as the larger aggregated number of subscribers/audience adds to the value of advertising spots/space and a conglomerate's ability to offer cross-platform distribution systems for ad messages makes it a more efficient advertising choice. The nature of public goods, on the other hand, encourages the geographic/international diversification of content products, as the incremental costs are minimal for such expansions. Finally, because of the importance of cultural sensitivity and understanding of the regulatory environment, global media conglomerates are more inclined to diversify into related product/geographic markets to take advantage of the acquired local knowledge and relationships.[1] The dependency on local communication/media infrastructure may also lead to a diversification strategy that is geographically related (i.e., regionalized), as geographically clustered countries are often at similar stages of infrastructure development and clusters of media distribution systems may lead to cost/resource-sharing benefits.

3 Examining Media Diversification in the Mobile Wireless Sector

Based on the notions of relatedness and complementary resource alignment as well as the incentive for international expansion into other media markets, we expect that a media conglomerate would have little interest in the wireless sector if product relatedness were its dominant diversification strategy. On the other hand, a media conglomerate would be more interested in the wireless sector if there were established complementary resources for it to utilize. By the same token, we would see more diversification activities into the wireless sector if there were a strong presence in related regions (especially the regions with strong wireless demands).

An exploratory case-study method was adopted for the diversification analysis in this chapter. As suggested by previous researchers, case study is most appropriate when a case represents a special set of circumstances that warrant in-depth investigation (Bradshaw & Wallace, 1991; Tellis, 1997). Researchers examining transnational media management have frequently used case studies, which provide more in-depth reviews of the evolution of transnational strategy and operations (Hollifield, 2001).

One of the most important steps for a case study is the selection of cases that provide insight to the phenomenon to be examined (Yin, 1993). We selected the top seven global media conglomerates based on their overall revenues in year 2000-2001 for comparative examinations because of their market leadership role in the industry. Data for the conglomerates' geographical/product diversification and resources were collected through archival sources such as company annual reports, various financial resources such as Hoovers, Moody's, OneSource, SDC Platinum Mergers and Acquisition Database, and Gale Group Business Databases. We also reviewed the significant developments for each conglomerate in the last three years as reported in financial trades and included in the Gale and OneSource databases to assess the conglomerates' strategic patterns that involved or might lead to diversification activities in the wireless sector.

To assess the degree of product diversification, we reviewed the number of business units and sectors (by SIC codes) present for each conglomerate. We also studied the M&A history of the conglomerates as recorded in the SDC Platinum Mergers and Acquisition Database published by *Thompson Finan-*

cial Securities Data. Specifically, we examined the M&A transactions in the last ten years (1992-2002) involving the seven media conglomerates in the wireless and wire-line telecommunications sectors.[2] To measure the extent or multiplicity of foreign markets in which a media conglomerate operates, we investigated the numbers of countries the conglomerates entered during the last ten years in their pursuits of M&A transactions (as an acquirer).[3] We further reviewed the M&A transactions occurring during the period in each region to investigate the core regions of international diversification for each conglomerate.[4]

4 State of International Product Diversification of the Leading Media Conglomerates

There is a range of product and international diversification among the leading conglomerates based on our analysis of the conglomerates' business units, sector presence, and recent M&A activities (see Figure 1). In terms of product diversification, the European Vivendi Universal and Bertelsmann had most diversity, whereas Viacom and News Corp. were in the least diversified group that relied heavily on advertising revenues and was most aggressive in pursuing the product relatedness strategy. It seems that the conglomerates are more motivated to approach related product diversification when there is a need to develop more attractive advertising opportunities (e.g., cross-platform advertising). In essence, the dual-revenue source characteristic of the media products might have influenced the strategic direction of a media conglomerate. Overall, based on the reviews of media sector presence and revenue contributions, one can conclude that Time Warner was the most diversified conglomerate in the global "multimedia" marketplace, followed by Bertelsmann, Viacom, and Vivendi Universal (the former was more distribution/outlets diversified, while the latter was more content diversified), Disney, News Corp., and lastly, Sony.

In regard to international diversification, partially due to the importance of the North American media markets, the most geographic diversifiers were non-U.S. corporations such as Vivendi Universal, Bertelsmann, and Sony according to their international M&A activities and foreign revenue dependency (see Table 1).[5] While Vivendi Universal and Bertelsmann have adopted a less related geographic diversification approach, Disney and, to a lesser degree, Viacom have preferred geographic relatedness. Time Warner was in a class by itself, taking more a middle-of-the-road approach.

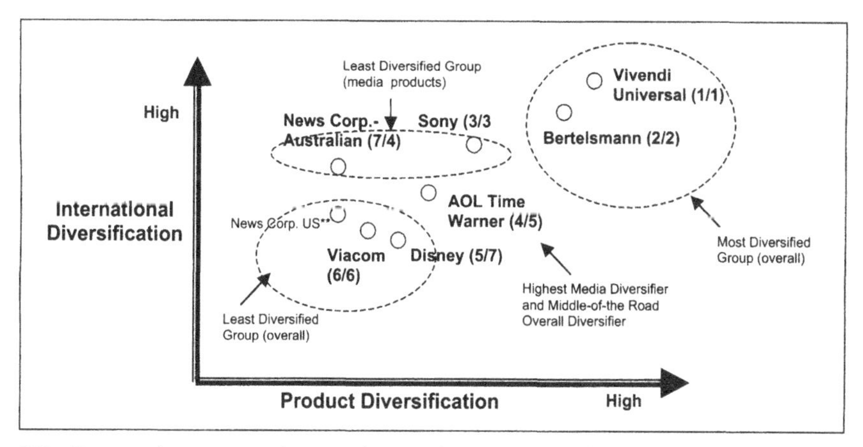

*The first number in parenthesis is the conglomerate's product diversity ranking, while the second is its geographic diversity ranking.
**News Corp. would become less internationally diversified if it were treated as a U.S. firm.

Figure 1: Relative International Product Diversification of Global Media Conglomerates*

An examination of the regional diversification patterns based on the conglomerates' M&A activities reveals the essentiality of the North American region (see Table 1). While Vivendi Universal and Bertelsmann focused on the Western European region, Sony and News Corp. concentrated in their home region, Asia Pacific. As Disney and Viacom chose to stay close to home, Viacom attempted to diversify into Western Europe, and Disney tested the Latin American markets. Time Warner, on the other hand, is competing directly against the two European conglomerates, with a focus on the North American and Western European regions. The uneven distribution of M&A activities between regions during the last ten years was consistent with the previous proposition that regional experience and relationships are best realized in "related" international diversification.

Note that Western Europe is especially important as a wireless market currently in comparison to the North American region; in this case, even though Vivendi and Bertelsmann have lesser degrees of geographical relatedness, they still have a very strong presence in a region that's very important in regard to the wireless sector.

Table 1: International Diversification of Global Media Conglomerates by Regions According to M&A Activities

Region	Vivendi Universal	Bertelsmann	Sony	News Corp.	Time Warner	Viacom	Disney
Countries Entered as an Acquirer	36	28	20	18	22	20	11
Regions Entered as an Acquirer	7	6	5	4	6	7	4
Asia Pacific	12	8	46	71	8	3	2
Western Europe	194	185	13	24	38	14	4
Central & Eastern Europe	30	8	1	3	4	1	0
North America	81	41	36	90	89	63	51
Latin America	3	2	2	0	1	4	6
Central America/the Caribbean	2	0	0	0	0	1	0
Near East	3	1	0	0	1	0	0
Africa	0	0	0	0	0	2	0

Source: SDC Platinum M&A Database

5 Conglomerate Diversification into the Telecommunications Sector

We also examined specifically the conglomerates' mergers and acquisitions of wireless telecommunications firms to assess the media's interest in the wireless business. We reviewed each conglomerate's M&A transactions targeted at the firms in the wireless sector (SIC 4812) during the same ten-year period (see Table 2). Overall, Vivendi Universal, and News Corp.,[6] followed by Time Warner, were relatively more active than the rest of the conglomerates. Bertelsmann, Disney, and Viacom had no wireless telecommunications-related M&As during this period. By comparison, Vivendi Universal was also more aggressive than the other two in diversifying into the wireless sector "internationally" (i.e., 10 of the 10 transactions were international).

6 Wireless Activities of the Leading Media Conglomerates

Though we have not observed an overwhelming interest in the wireless sector for these media conglomerates as they attempt to diversify through equity acquisitions, many of the conglomerates have participated in the mobile wireless sector via various types of non-equity alliances with firms that provide either the conduit, equipment, or content for wireless services. We will now discuss some of these conglomerate activities involving the wireless sector.

As evidenced by the M&A analysis, Time Warner is a relatively active wireless player among the leading conglomerates. In fact, one of Time Warner's

Table 2: Global Media Conglomerates' M&A Transactions Targeted at Wireless Firms 1992-2002

	Vivendi Universal	Bertelsmann	News Corp.	Time Warner	Sony	Viacom	Disney
Wireless (4812)	10	0	10	6	2	0	0
International	10	0	3	2	1	0	0
Domestic	0	0	7	4	1	0	0

Source: SDC Platinum M&A Database

Table 3: Selected Wireless Partners and Activities of the Leading Global Media Conglomerates

Time Warner		
AOL Mail/Instant Messenger Agreements with Wireless Firms – AT&T Wireless – Motorola – Deutsche Telecom AG's VoiceStream Wireless – Aether Systems – VoiceStream Wireless – Psion – Genie (a European mobile Internet player) – Nokia – Research in Motion – Bellsouth – Arch Communications – Sprint PCS – NTT DoCoMo	*Content-Related Agreements* – Motorola – Warner Bros. – AT&T Digital PocketNet – OmniSky	*Licensing Agreements of Text Input Software* – Hitach – China Kejian Corporation Ltd – Arima – Sendo – Toshiba – Hyundai Electronics – Sony – Telit Mobile Terminal (Italy-based mobile phone maker) – Panasonic
Disney		
Content-Related Agreements – Telenor Mobile – SK Telecom (a Korean mobile Internet service provider) – AT&T Wireless – Sprint PCS – Deutsche Telecom AG – Taiwan Celluar Corporation		
Bertelsmann — BeMobile		
Content-Related Agreements – Quam and Bertelsmann's BeMobile – Isyndicate – Tera Lycos – AOL	*Software/Platform Development* – Zap Business Communication Systems (JuniorNet B-to-B Wireless Access)	

(continued)

Sony	
Handset and Software/Platform Development – Ericsson – Nokia – Sun Microsystems – Microsoft	
Vivendi Universal — Moviso	
Multimedia/Wireless Platform Development – Thomson multimedia – Premium Wireless – Vodafone Group Plc – Vizzavi Europe	
News Corp.	
Content-Related Agreements – OmniSky International	*Mobile Internet Services* – Singapore Telecom
Viacom	
Content/Brand-Related Agreements – nGame	*Mobile Internet/Messaging Services* – OmniSky International – IBM – VibesMedia

Sources: OneSource.com, company press releases and websites, & Hoovers.com.

core businesses, Internet access services, is strongly tied to the development of wireless web. A review of Time Warner's wireless activities revealed that AOL has been focusing on expanding its e-mail and instant messenger services and on licensing the T9 Text Input software to mobile-device makers and wireless service providers (see Table 3). Relatively, it has focused on improving the wireless accessibility of its Internet core product and not attempted to capitalize on its content properties and brands as its counterparts such as Disney and Bertelsmann have (see Table 4). In fact, Disney, with a high relatedness diversification strategy, has expanded to the wireless sector by emphasizing primarily the transfer of its branded media content (such as images/characters and short program content) to the wireless outlets. Bertelsmann, with low overall relatedness but strong media brands and an European presence, has also been active in forming alliances to enable the wireless transfer of its branded content products.

As a wireless device maker, Sony has, on the other hand, emphasized technical alliances or licensing agreements with other wireless-device makers such as Ericsson or Nokia and software companies such as Sun Microsystems and Microsoft. In essence, with relatively less relatedness and strong media brands, Sony seems to focus on improving wireless accessibility through the development of competitive, seamless wireless devices and software. As a highly diversified conglomerate, similar to its counterparts with relatively less product relatedness, Vivendi Universal has centered its wireless efforts on improving the accessibility of contents. Its Moviso subsidiary has formed alliances

Table 4: Strategic Focus and Relatedness/Resource Alignment of the Global Media Conglomerates

CONGLOMERATE	STRATEGIC FOCUS	STATE OF DIVERSITY/RELATEDNESS
Time Warner	Accessibility (wireless access of Internet products)	– Moderate product relatedness (low media relatedness; core Internet product) – Moderate regional relatedness
Sony	Accessibility (wireless device & platform development)	– Moderate product relatedness (high media relatedness) – Moderate regional relatedness (strong Asia Pacific presence)
Vivendi Universal	Accessibility (Platform development)	– Low product relatedness – Low regional relatedness (strong European presence)
Bertelsmann	Transfer of media content/brands	– Low product relatedness (strong media brands) – Low regional relatedness (strong European presence)
Disney	Transfer of media content/brands	– High product relatedness – High regional relatedness
News Corp.	Limited content & accessibility	– High product relatedness – Moderate regional relatedness
Viacom	Limited content & accessibility	– High product relatedness – High regional relatedness

to develop the smooth transfer of content products using multimedia/wireless platforms. Finally, News Corp. and Viacom, the two conglomerates that have employed a most related diversification strategy, have initiated limited partnerships with wireless firms in areas of content for mobile wireless users and Internet access/messaging services (see Table 4). Our observations of these conglomerates' wireless activities seem to be consistent with our previous propositions on the importance of relatedness and resource alignment.

7 Strategic Patterns of the Leading Media Conglomerates Concerning the Wireless Sector

A review of the significant developments in the wireless sector in the last few years involving the selected global media conglomerates revealed some interesting trends that paint a more descriptive picture of the wireless diversification efforts of the conglomerates.

7.1 Competitive-Cooperative Relationships Between Leading Global Conglomerates

An interesting phenomenon that we have observed in the last few years is the interdependency that turns the leading global media conglomerates not only into competitors but also into partners (Chan-Olmsted, 2004). For example, in an attempt to expand to the European market, Time Warner allied with Vivendi Universal under an agreement in which Vivendi Universal group companies exchanged shares with AOL's European holdings and entered into other distribution and marketing agreements with Time Warner. As Sony and Time Warner formed various partnerships to develop home networking technologies that provide a variety of consumer content and services for a broadband environment, News Corp. and Vivendi Universal entered a worldwide co-publishing agreement, which grants Vivendi Universal exclusive rights to manufacture, market, and distribute certain News Corp. content products. Disney and News Corp. have established a joint venture that offers a new broadband entertainment service called Movies.com, which provides movies and other entertainment content on demand to U.S. consumers. Recently, Bertelsmann and Time Warner co-invest in a mobile gaming and entertainment company, Codeonline, to develop mobile services for brands/content such as "Who Wants To Be A Millionaire?," "Trivial Pursuit," and "E.T."

7.2 Alliance with Wireless Partners to Improve Content Accessibility and Internet Presence

While the global media conglomerates do not aggressively "diversify" into the wireless sector, they have actively sought strategic alliances with wireless firms to ensure the accessibility of their "content" products via the wireless platform and to develop a presence in wireless Internet services. For example, Vivendi Universal has invested in many U.S.-based wireless companies such as Premium Wireless Services and the satellite television service EchoStar. In Europe, working with Thomson Multimedia, Vivendi Universal is testing multimedia uses of its content for mobile phone, in addition to its agreement with Vodafone in establishing a new 50/50 Internet company to develop and operate a branded Multi Access aiming at developing the wireless Internet.

As for other global conglomerates, Disney has signed distribution agreements with Telenor Mobile for access to Norway, Sweden, Denmark, and Finland users, and with SK Telecom for access to Korean users, in addition to its agreements with the American wireless players, AT&T and Sprint PCS. Time Warner continues to develop partnerships with a variety of wireless companies such as NTT DoCoMo, Deutsche Telecom AG's VoiceStream, Motorola, AT&T, OmniSky, Sprint PCS, Psion, and Aether Systems, Inc., to increase its presence and accessibility in the wireless market. Again, as Sony allied with various wireless companies such as Nokia to develop an open and common platform for wireless services, News Corp and OmniSky, a provider of comprehensive branded wireless Internet service for handheld mobile devices, formed a joint venture to explore international opportunities for wireless Internet services.

7.3 Position for the International Distribution of Mobile Interactive Television Services

To ensure that they are at the forefront of the Internet-driven wireless broadband television revolution, global media conglomerates are establishing interactive television and Internet content and services and are allying with the firms that facilitate such services. For example, Time Warner has allied with Cisneros to produce online and television content to tap into the Internet growth in Latin America (Mermigas, 2001). Time Warner has also formed joint ventures with Microsoft to package Miscrosoft's Internet audio and video technology with AOL's Internet service. Disney's Internet Group and BellSouth have joined an agreement under which Disney licenses selected online content on a non-exclusive basis to BellSouth for re-distribution via the Bell-

South Internet Service portal. Viacom, via its CBS holding, has allied with Microsoft to deliver interactive television programming. Sony and RealNetworks have formed a strategic alliance under which the two adopt each other's technology to co-market products. Sony is also working with Microsoft to fund development of an interactive Microsoft television concept. While Vivendi, Canal+, and Vodafone AirTouch formed a joint venture that created a company to provide a multi-access Internet portal for Europe, News Corp. entered a partnership with Worldgate to develop interactive television services and again allied with GigaMedia to develop interactive television in Asia. Most of these Internet-based interactive television ventures include plans to distribute interactive television services via both wired and wireless platforms.

8 Drivers for Mobile Wireless Diversification

By nature, for the global media conglomerates, the decision to diversify is a matter of degree and target and not the decision of whether to diversify (Compaine, 2001; Noam, 2005). Many media trade publications have identified up to 50 conglomerates that are actively pursuing a diversification strategy in the global media marketplace (Global Top 50, Aug 27, 2001; U.S. Top 100, August 20, 2001). We proposed that, in the case of wireless diversification for a media corporation, as prescribed by the industrial economics perspective of diversification which stresses the importance of external environment in shaping the strategic behavior of a firm, the general environment of a target country such as its regulatory, economic, technological, cultural, and social (e.g., education) environment influences not only the attractiveness and characteristics of the wireless market in that country but also another set of important country specific external factors—the country's wireless communications infrastructure and demand for the wireless products (see Figure 2). These environmental factors also directly impact the attractiveness of the wireless industry in that country. Continuing on the industry economics theory of diversification, a media conglomerate's decision to enter the wireless market is likely to be determined by its target industry's basic wireless market characteristics such as market size, growth rate, profitability, and competition, as well as the factors of product/geographical relatedness and content-distribution complementary alignment as discussed previously.

Subscribing to a resource-based view of strategy, we believe that in addition to many internal/resource drivers such as financial performance, current diversifying equity-based holdings, internationalization expertise, and mar-

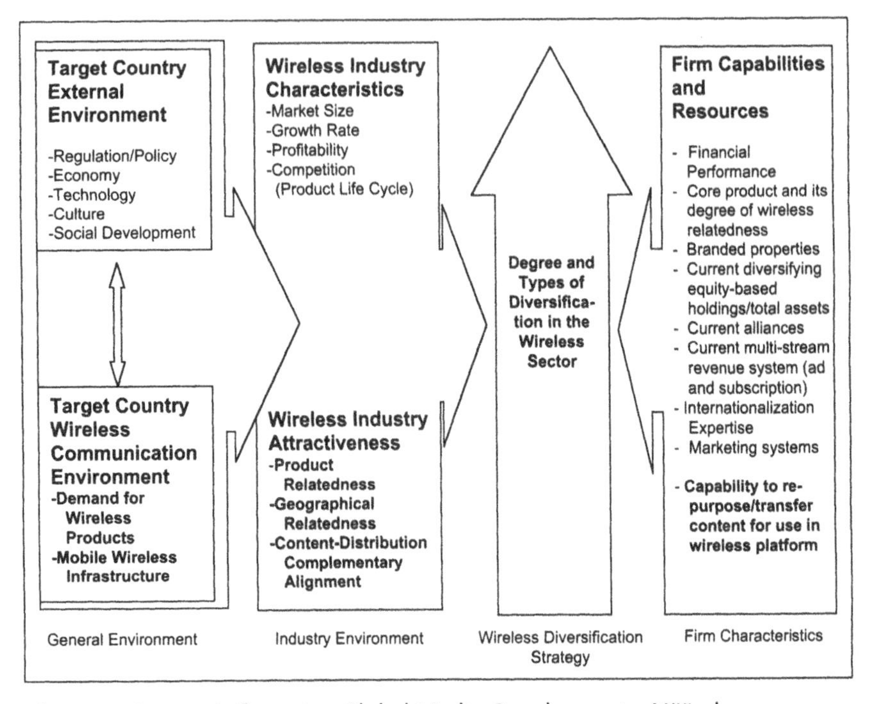

Figure 2: Factors Influencing Global Media Conglomerates' Wireless Diversification Strategy

keting systems, that have been established to impact corporate diversification in previous literature (Miller & Shamsie, 1996; Barney, 1991; Grant, 1991), many media specific resources affect a conglomerate's diversification decision. For example, a media corporation's existing alliances with other wireless firms, its dependency on and wireless relatedness of the core product, and branded properties would shape the conglomerate's preference in both product and geographic diversification in the wireless sector. Most importantly, a conglomerate's capability of transferring or re-purposing content products for the wireless outlets as well as the availability of a multi-stream revenue system would also determine the degree of geographic diversity and the extent, directions, and mode of product diversification into the wireless market (see Figure 2).

9 Conclusion

The global media conglomerates have approached the mobile wireless medium with limited asset diversification but numerous strategic alliances to improve the wireless accessibility of their media content/brands and Internet services. The factor of product relatedness apparently presents an obstacle for aggressive extension to the wireless sector by the media conglomerates. Complementary resource alignment also seems to be a pre-requisite for wireless diversification as related geographical markets appears to be a preference for the conglomerates. We also found the necessity of owning North American media assets, especially those of content properties; the importance of allying with partners that improve content accessibility globally; and the need to explore the new media opportunities via alliances with international media facilitators, distributors, and content producers. There is an observable oligopolistic behavior between the leading conglomerates as these competitors frequently become collaborators for business ventures in a less certain market environment (e.g., new geographic or product markets).

Subscribing to both the industrial economics and resource-based view of strategy, we believe that the demand and infrastructure in a target country, the country's wireless industry's attractiveness in regards to product relatedness and other factors and a conglomerate's ability to repurpose its branded content for the wireless platform in that country will remain the primary drivers that will determine the global media conglomerate's participation on the wireless sector.

In conclusion, diversifying into the wireless industry with a goal to deliver mass media content products using the wireless platform is currently inconsistent with most media conglomerates' strategic approaches, namely, the considerations of relatedness, resource alignment, and the defensive tendency. It is more likely for these conglomerates to first form strategic alliances to ensure content accessibility and to develop a presence in wireless Internet services. This strategy requires minimum resource commitments and risks. As the conglomerates observe the growth of demand for wireless content and the maturing of wireless technology, they may choose to develop complementary resources and/or explore product relatedness in their next diversification move.

Endnotes

1. The geographic market relatedness may also be examined in terms of language and cultural relatedness (e.g., Spanish language media content).
2. Note that the premise of our examination is that the diversification entry approach is predominately through mergers and acquisition than internal development as documented by many studies. See Mermigas, 2001; Albarran and Chan-Olmsted, 1998; and Shearer, 2000.
3. We decided to use this measure instead of the number of countries where a conglomerate has established operations because of the complex and inconsistent definitions for international branches each conglomerate has adopted, which may include subsidiaries as well as affiliates and nonaffiliated licensees, and the discrepancies in the numbers of reported countries entered by different divisions of each conglomerate.
4. The classification of the eight regions was based on the considerations of cultural, economic, and physical geographic divisions and adopted from the Economic Growth Regional Classification framework proposed by the Economic Growth Center of Yale University. See http://library.yale.edu/socsci/egcclass.html.
5. About 49% of Vivendi Universal's revenues were from foreign sources, 64% for Bertelsmann, 67% for Sony, 26/90% for News Corp. (depending whether treating it as an Australian or U.S. firm), and 21% for both Viacom and Disney.
6. The wireless investments that News Corp. made were mainly in the satellite television industry.

References

Ansoff, H. I. (1957). Strategies for diversification. *Harvard Business Review, 35*, 113-124.

Ansoff, H. I. (1958). A model for diversification. *Management Science, 4*, 392-414.

Barney, J. (1991). Firm resources and sustained competitive advantage. *Journal of Management* 17, 99-120.

Berry, C. H. (1975). *Corporate Growth and Diversification.* Princeton, NJ: Princeton University Press.

Booz, Alllen and Hamilton. (1985). *Diversification: A Survey of European Chief Executives.* New York, NY: Booz, Allen and Hamilton, Inc.

Bradshaw, Y. & Wallace, M. (1991). Information generality and explaining uniqueness: the place of case studies in comparative research. *International Journal of Comparative Sociology, 32*, 154-172.

Buhner, R. (1987). Assessing International Diversification of West German Corporations. *Strategic Management Journal, 8*, 25-37.

Cavaglia, S. M, Melas, G. D., & Tsouderos, G. (2000). Cross-industry and cross-country international equity diversification. *Journal of Investing, 9*, 65-71.

Caves, R. E. (1982). *Multinational enterprises and economic analysis.* Cambridge, NY Cambridge University Press.

Chandler, A. D. (1962). *Strategy and Structure: Chapters in the History of the American Industrial Enterprise.* Cambridge, MA: MIT Press.
Chan-Olmsted, S. M. (2004). In search of partnerships in a changing global media market: Trends and drivers of international strategic alliances. In Picard, R. G. (Ed.), *Strategic Responses to Media Market Changes,* Jonkoping, Sweden: Jonkoping International Business School.
Chan-Olmsted, S. & Albarran, A. (1998). The global media economic patterns and issues. In A. Albarran, & S. Chan-Olmsted (Eds.), *Global media economics: commercialization, concentration and integration of world media markets.* Ames, IA: Iowa State University Press.
Chatterjee, S. & Wernerfelt, B. (1991). The link between resources and type of diversification: theory and evidence. *Strategic Management Journal, 12,* 33-48.
Compaine, B. (2001). The myths of encroaching global media ownership. Retrieved August 26, 2004 from the World Wide Web: http://www1.primushost.com/~bcompain/WOTM/media_myths.htm.
Global Top 50. *Advertising Age,* Aug. 25, 2001.
Gort, M. (1962). *Diversification and Integration in American Industry.* Princeton, NJ: Princeton University Press.
Grant, R. M, & Jammine, A. P. (1988). Performance differences between the Wrigley/Rumelt Strategic Categories. *Strategic Management Journal, 9,* 333-346.
Grant, R. M. (1987). Multinationality and performance among British manufacturing companies. *Journal of International Business Studies, 18,* 79-89.
Grant, R. M. (1988). On dominant logic, relatedness and the link between diversity and performance. *Strategic Management Journal, 9,* 639-642.
Grant, R. M. (1991). The resource-based theory of competitive advantage. *California Management Review, 33,* 114-135.
Greenspan, R. (July 6, 2004). Global mobile population growing. *ClickZ Stats.*
Hisey, K., & Caves, R. E. (1985). Diversification strategy and choice of country: diversifying acquisitions abroad by U.S. multinationals, 1978-1980. *Journal of International Business Studies, 16,* 51-64.
Hitt, M. A., Hoskisson, R. E., & Ireland, R. D. (1994). A mid-range theory of the interaction effects of international and product diversification on innovation and performance. *Journal of Management, 20,* 297-326.
Hitt, M. A., Hoskisson, R. E., & Kim, H. (1997). International diversification: effects on innovation firm performance in product-diversified firms. *Academy of Management Journal, 40,* 767-798.
Hollifield, C. A. (2001). Crossing Borders: Media Management Research in a Transnational Market Environment. *Journal of Media Economics, 14,* 133-146.
Kim W. C., Hwang, P., & Burgers, W. P. (1993). Multinational's diversification and risk-return tradeoff. *Strategic Management Journal, 14,* 275-286.
Kobrin, S. J. (1991). An empirical analysis of the determinants of global integration. *Strategic Management Journal, 15,* 17-37.
Kogut, B. (1985). Designing global strategies: comparative and competitive value-xxx chains. Part 1. *Sloan Management Review, 27,* 15-28.
Madura, J. & Rose, L. C. (1987). Are product specialization and international diversification strategies compatible? *Management International Review, 27,* 38-44.
McChesney, R. W. (1999). The new global media. *The Nation, 269*(18). 11-15.

Mermigas, D. (2001). Bigger is better in a soft market. *Electronic Media, 20*(37), 30.
Michel, A., & Shaked, I. (1984). Does business diversification affect performance. *Finance Management, 3*, 18-25.
Miller, J. C., & Pras, B. (1980). The effects of multinational and export diversification on the profit stability of U.S. corporations. *Southern Economic Journal, 46*, 792-805.
Miller, D. & Shamsie, J. (1996). The resource-based view of the firm in two environments: The Hollywood film studios from 1936 to 1965. *Academy of Management Journal, 39*, 519-543.
Miller, D. & Shamsie, J. (1999). Strategic responses to three kinds of uncertainty: product line simplicity at the Hollywood film studios. *Journal of Management, 25*, 97-116.
Montgomery, C. A. (1982). The measurement of firm diversification: Some new empirical evidence. *Academy of Management Journal, 25*, 299-307.
Noam, E. (2005). *Ownership and Concentration in the US Communications Industry*. New York: Oxford University Press (forthcoming).
Palepu, K. (1985). Diversification strategy, profit performance, and the entropy measure. *Strategic Management Journal, 6*, 239-255.
Qian, G. (1997). Assessing product-market diversification of U.S. firms. *Management International Review, 37*, 127-149.
Rumelt, R. (1984). Toward a strategic theory of the firm. In Lamb, R. (Eds.), *Competitive strategic management* (pp. 556-570). Englewood Cliffs, NJ: Prentice-Hall.
Salter, M. S., & Weinhold, W. S. (1979). *Diversification Through Acquisition*. New York, NY: Free Press.
Sambharya, R. B. (1995). The combined effect of international diversification and product diversification strategies on the performance of U.S.-based multinational corporations. *Management International Review, 35*, 197-218.
Tallman, S. & Li, J. (1996). Effects of international diversity and product diversity on the performance of multinational firms. *Academy of management Journal, 39*, 179-196.
Tellis, W. (1997). Introduction to case study. *The Qualitative Report* [e-journal], *3*(2). Retrieved from the World Wide Web: http://www.nova.edu/ssss/QR/QR3-2/tellisl.html.
U.S. Top 100. *Advertising Age*. Aug. 20, 2000.
Williamson, O. E. (1981). The modern corporation: origins, evolution, attributes. *Journal of economic literature, 19*, 1537-1568.
Yin, R. (1993). *Application of case study research*. Beverly Hill, CA: Sage.

IV

Policy Models

14
Exclusive Rights in Information and Mobile Wireless Mass Media

*Yochai Benkler**

1 Introduction

I had some misgiving when I was invited to write a chapter on the implications of exclusive rights in information (ERIs; more commonly known by the less analytically neutral term "intellectual property rights", or IPRs) for the future of delivery of mass media content to mobile wireless devices. This is, to some extent, a futile project, as I do not actually think that the correct way for mobile wireless data communication to go is a replication of the mass media model. Quite the contrary, that is a model that I have spent a good deal of writing arguing is economically inefficient and normatively unattractive (Benkler, 1998, 2001, 2004). What I propose in these pages, therefore, is to provide a descriptive account of the ERI implications for any business model that could seriously be called mass media delivery to mobile wireless. I will then sketch my objections to such a business model, suggesting instead that the core of mobile wireless communications should be, and will be, the provision of high-speed mobile Internet access through equipment that will enable license-free, network-owner-independent communications, not mass media. Adopting such a model will render the issues that are of concern to a mass media model irrelevant to mobile wireless.

2 Rent-Seeking Politics, Not Rational Policy

As we look at the different regimes for copyright licensing and royalty regimes, we see that each new technological development offers a new battlefield. The owners of copyright operate from the presumption, reasonable from their perspective, that they ought to get as high a return as possible from every new market segment opened up by a new technology of transmission. Carriers, service providers, and equipment manufacturers, on the other hand, want to maximize the value of whatever system they are selling by using it to transmit as much content as possible, at as low a price as they can. The result has been a series of battles ranging from the introduction of player pianos to radio, television, cable, satellite, and the Web. There is no unifying principle behind the resolution of these, no single template. There is a general range of possible outcomes, ranging from "copyright holder controls transmission at will" to "copyright holder has no right", but often settling on some form of compulsory license with some level of royalty. The contours of the license requirement and the rate of the royalty are generally determined by the political jockeying rather than by specific application of general principles.

2.1 Parameters of Constraint

2.1.1 Two Ideal Type Business Models

To motivate the analysis, let us adopt two opposite visions of mobile wireless delivery of mass media content. At one end of the spectrum is a business model that may be conceived of as the *"Glorified Car Radio" (GCR)*—mobile devices available purely for push content delivery selected by the content provider. Interactivity is minimized. Similarly, adaptability and functionality are constrained. The GCR model focuses the role of the consumer as a largely passive recipient of information, whose role in the system as payor, rather than a source of value into the system, except where the recipient him—or herself is the product sold to advertisers.

At the other end of the spectrum are business models that provide a form of high-speed mobile Internet access, whether in an owned network or on an open wireless network that is ownerless, and built purely of collaborating end-user devices (Werbach, 2001; Benkler, 2002, 2004). The most visible current instantiation of this model are the examples of 802.11x family of standards, or WiFi networks, described by other chapters in this volume, and can also be seen in various proposals to enhance the currently limited web access and e-mail presently incorporated into most cellular devices. As handheld devices become increasingly integrated and PC's more seamlessly wireless the line between handsets and PCs begins to blur. "Equipment" here correlates with the wide-spanning functionality and adaptability of the desktop PC—encouraging a variety of uses, the transferability of content, and the expansion of applications—and the mobility of handsets. Equipment and services are built flexibly to adapt to the needs of the user, rather than optimized to deliver content selected by a provider to the user. Networks are designed to be network owner-independent.

This *Mobile High-Speed Internet Access (MHIA)* business model lends itself to a variety of services, both push and pull. Content may take advantage of the location-oriented needs of mobile consumers. Services may also provide the entertainment content delivered via traditional mass media mechanisms. Critically, however, such proprietary "mass media" content is to be accessed and delivered with equal weight alongside non-proprietary content and personalized interactions. In its pure form, then, this model puts a premium on the characteristics we find in dial-up access to an open web: a user-controlled environment that promotes, rather than curtails, interactivity, and choice. One obvious impure form would be a model that offers Internet access, but introduces constraints on the flexibility afforded to users at the equipment-end in order to privilege certain content—say, affiliated content—through preferential delivery. That is to say, an impure model is aimed at evoking the potential pressures from the mass media model on the Internet model that one sees presented in the open access debate over cable broadband.

The general layout of the constraints that the law will place on different providers of wireless mobile communications will largely depend on how close a given business model is to one or the other of these pictures. These will inform and shape the type of constraints that courts and legislatures are likely to place on providers.

2.2 Core Determinants of the Shape of Regulatory Intervention

There are three elements to mobile wireless delivery, which in combination will lead a business model to look, from the copyright perspective, more like a GCR or more like a MHIA service. These are (1) the end user device, (2) the transmission infrastructure, and (3) the content. One can organize the relationship among these three components in a number of ways, corresponding to the extent to which each component is tied to another and owned by a service provider.

The transmission service provider either controls the device or it does not.

The infrastructure owner either controls the delivery service (dedicated ISP or cable services, etc.), or it does not (open access or common carriage interconnection).

The device either controls the content (strongly through hardware limitations or weakly through startup defaults), or it does not.

The infrastructure owner either controls the content delivered (completely as in TV or weakly as in default gateways), or it does not.

While obviously it is analytically possible to combine these parameters into more combinations, it is sufficient to consider the outlines of a number of prototypical business models that need to be considered from a policy/law perspective given these characteristics:

Radio and Television Terrestrial Broadcast model: The device is dedicated to the infrastructure service, the infrastructure is dedicated to use by one or a set number of providers, the service provider controls the content on its service.

Cable Broadband ISP: The device is dedicated to the infrastructure service, the infrastructure is dedicated to use by one service provider, the content is shaped by the infrastructure service through gateway defaults, but is otherwise on an Internet-pull model.

Dial-up ISP model: The device is not dedicated to the infrastructure service, but rather can be used to connect to any infrastructure service within range, the infrastructure is not dedicated to one service provider, and the content is on a pull-Internet model.

Mobile DVD players: The device is dedicated to a particular type of content; the infrastructure is competitive and not dedicated to the content (that is, Walmart and Tower Records can compete on distribution).

Bloomberg Radio (modified): As a sales gimmick, Bloomberg Media destributed free little AM radios that could only be tuned to one station—Bloomberg Radio. This, in modified form, stands for the model where the end-user hardware is hardwired, or at least its software is dedicated, to specific content

- which can be transported over infrastructure owned by the hardware and content vendor
- or can be devised so that it will always deliver stated content irrespective the transmission infrastructure to which it is connected.

Models that will operate closer, along the spectrum of transmission models, to the GCR model, will operate in a space defined by the permission/compulsory license/royalty set of concerns. The baseline question for any commercial business model built around pushing commercially produced mass media content to mobile consumers is, therefore, whether it needs to negotiate a license with the copyright owners, or whether it can fall under some framework for a compulsory license with a statutorily-fixed royalty.

The conclusion with regard to the general category of GCRs—that is to say, devices that push content to consumers over infrastructure owned by the programmer—is that it is possible that the transmissions will be deemed to have fallen under a category of service already covered by a compulsory license, in which case one would want most to be treated like terrestrial broadcasters carrying audio, or like cable companies retransmitting broadcasts. The terrain is treacherous however. The basic approach of the content industries, quite justifiably from their perspective, is to fight any new technology of retransmission or broadcast as though it were a new battle, where the other side is not as strong as the incumbent terrestrial broadcasters, so as to limit the compulsory license to a bare minimum and to raise the royalty to a maximum. The same is likely to apply to mobile wireless delivery, were licensees could in principle try to claim to be FCC licensed "terrestrial broadcasters," but that would be a very weak argument unlikely to be accepted by courts. More likely, a new regulatory proceeding to determine the appropriate licensing fee would be required.

The set of constraints and considerations are different when the wireless service provider moves further away from the glorified car radio model, and toward a mobile high-speed Internet access model. Here the framework of the newly emerging technology control laws pushed by the copyright industries—mostly the DMCA and other modes of ISP or service provider contributory liability theories—as well as the looming hardware design control statute—the CBDTPA—are most relevant. The more the infrastructure is available for users to pull and exchange content of their own choosing, however, the further away we get from something that could properly be called "mass media" delivery of content.

The baseline concern for liability in the MHIA model is the background contributory liability from copyright law. If the users use the provider's facilities in ways that violate copyright, there is some concern that this will result in contributory liability suits. In the early 1990s there were a number of suits in copyright as well as other contexts on this theory against ISPs. The paradigm case, though, is not a service case but an equipment case, in which the movie industry sued Sony to stop distribution of VCRs. The suit failed, on a theory that the VCR had a substantial noninfringing use, to wit, allowing users to tape television programs that were broadcast when they were not at home, so they could watch them at a more convenient time. The case was later applied at an appeals-court level to permit mp3 players, against a structurally similar suit by the recording industry. These cases suggest that any generally usable high-speed service, not dedicated in some important way to copying and distributing copyrighted materials, will be safe.

Above this baseline level of potential liability, there are specific arrangements introduced by the DMCA. Here, a true exemption from liability as an ISP applies to providers who are purely "carriers." But as the service bundles more value added services, like a portal, search and indexing functions, and so on, other conditions apply, primarily the notice-and-takedown approach. That is, if the content provider knows of infringing content that is flowing over the ISP's system, and notifies the provider of the content, the ISP has to remove the infringing content. This was all well and good for a framework that depended on ISPs to store the delivered content, so there was always an ISP that had actual control over the storage space from which the offending materials were uploaded. The introduction of peer-to-peer technologies after the passage of the DMCA threw something of a monkey wrench into this mechanism. Napster was the example of a service provider that matched users to each other, but had no control over the content itself, and found it practically impossible to block the offending traffic.

This is not at all to say that all ISPs over which peer-to-peer occurs will be liable like Napster. A service provider that simply provides infrastructure, as mentioned, is immune even to notice and takedown requirements. But as value added services are added, as capacities to monitor traffic and control it increase, the service provider approaches liability for infringing materials carried over its network, even when it neither originates the content nor facilitates its delivery in any specific manner, other than offering general location devices and transmission capability. Failure to practicably implement a mechanism to control users' use led, at a practical level, to the closure of Napster.

The overall policy conclusion is that the further away high-speed mobile data services move from pure ISP service, the more they are engaged in controlling the information flow to their consumers, the more likely they are to be liable under copyright, and the more they will need to engage in the analysis I described for the glorified car radio model.

But this is only one dimension of the pressures that mobile providers of the high-speed data access model will face. There is pending legislation, intended to supplement requirements initially introduced in the DMCA to prohibit providers from giving consumers facilities that would allow them to circumvent digital rights management schemes (DRM). The proposed legislation takes the DMCA's approach of legally aiding vendors to control how users use their digitized products by regulating technology to its next logical step, by requiring equipment to be designed according to specifications that would assure its compliance with the copyright industries' standard for protecting their products. These standards are to be created by a body composed of copyright industry representatives and equipment manufacturers, but if they fail to reach agreement, the law leaves it to the FCC to set these standards for how personal computers and all other electronic devices are manufactured to fit the technological protection standards of the copyright industries. Should this legislation pass, mobile wireless providers who are focusing on data delivery will likely need to design their services and their hardware to comply with standards set by the government for purposes of copy control. This may be focused on the hardware manufacturers, but may also extend to designing the transmission mechanisms so as to recognize and track permissions. It is too early to speculate on the contours of such requirements, should they pass. The basic point to understand is that the domain of para-copyrights—various statutory requirements intended to build a buffer zone around the copyright industries to protect them from the vicissitudes of digitization—can have substantial constraining effects on the design and implementation of mobile high-speed wireless data access generally.

2.3 But this is all the Wrong Set of Questions ...

Most of the discussion up to this point has assumed that the primary person asking the question of "what are the copyright implications of mobile wireless mass media delivery" is someone who owns a license to some form of mobile wireless service, who is attempting to move from a model of mostly voice communications and maybe SMS, to a more mass media, bandwidth hungry, high value added service. This picture is wrong in a number of fundamental ways:

The model of data delivery in the future should not be built on licensed spectrum, but on unlicensed spectrum, like WiFi networks and other equipment in the 802.11 set of standards.

There will remain room for mobile license services, but mostly in latency-sensitive services like voice and real time video conferencing, not in less latency-insensitive services of the type that can now be delivered over high speed Internet connections, even including video on demand with a short latency period for buffering.

The central value in communication is human interaction, not mass consumption of finished goods. Business models for mobile communications should focus on putting tools in the hands of individuals to talk, make their own movies, play games with others, write their own restaurant guides, give directions to others on the street next to them, and so forth, not on pushing finished cultural goods at consumers.

If true, the better business models are less concerned with mass media delivery and its copyright-based limitations, and would move away from the model of control over infrastructure and the content it delivers in two directions. First, from license-based service toward end-user equipment that enables users to make flexible uses over time, and second, into the Internet, towards businesses intended as tools for users who are not constrained by the infrastructure and the device to select from a relatively sticky set of service options, as mobile phone users are. Apple's "rip, mix, burn, it's your music" advertising campaign is a particularly crisp example of the equipment move, where the mass media content is seen as simply an input, and the value is in the manipulation capacities that the equipment provides the users. Less in-your-face models, more directly relevant to mobile wireless communications, are Intel's push to introduce a radio on a chip, which is WiFi-based, or products based on proprietary standards for communications over unlicensed frequencies, like Motorola Canopy and Nokia Rooftops. An example of a company moving from mass media deep into the Internet cloud to form tools usable by a users with any device that enables Internet access is USA Networks, which seems to be gradually migrating from being primarily a mass media provider to being a major provider of tools for users to manage their own lives from anywhere—like Expedia.com, ticketmaster.com, and citysearch.com.

In the face of the possibility of technical and business model alternatives to the model of "spectrum licensee controls infrastructure and the content on it", it becomes possible and important to consider two important problems, from the perspective of public policy, with that model. The simple problem is the effect an infrastructure-owner-controls-content model has on competition in content and services. This is a problem we know well from cable broadband, and from the emerging parallel debate about the openness of DSL. This is a problem discussed in Eli Noam's piece in this volume, to which he offers an elegant solution.

The second, more intractable, difficulty for the "spectrum licensee owns infrastructure and controls content" model is that spectrum licensing and ownership are not a desirable regulatory structure for managing wireless communications (Noam, 1998). The full exposition is too long to be given here. The encapsulated argument is that the rapidly declining cost of computation has changed the efficiencies that can be gained by bandwidth management through exclusive rights—be they licenses or property rights—as compared to unlicensed or commons-based approaches. More capacity in wireless communications networks can be gained and its growth could be faster by improving the processing capabilities of transmitters and receivers and the intelligence of wireless communications networks design than by implementing a market in spectrum licenses.

The extent to which a transmitter can reach a receiver in a way that permits the receiver to differentiate between the transmitter's signal and other sources of radiation is dependent not only on bandwidth and power, but on some combination of the bandwidth and power of the signal sent, the processing power of the receiver and the transmitter, antenna design, and the network architecture of wireless devices. When processing was expensive and receivers had to be computationally simple to be affordable, transmission power and bandwidth were the sole variables that could be used to separate signal from noise. These, in turn, were indeed the focus of licensing, and later of auctioning, and in the spectrum-property market ideal, of property rights. The drop in the price of computation, and improvements in network architecture and in the potential of cooperation among wireless devices to increase the capacity of wireless networks, have now made it possible to model and solve the problem of allowing many users in a system to communicate without a wire without relying on exclusive control of any given channel. Property rights in spectrum impose transaction costs and limit the total bandwidth usable by devices, and the flexibility with which it can be used, relative to what wideband radios and software defined radios could use efficiently. Because of the limits they place on efficient utilization of intelligent end-user equipment, property-based wireless systems will have less capacity than, and grow capacity more slowly than unlicensed- or commons-based open wireless networks.

To hitch the wagon of efficient wireless communications to the dynamics of Moore's Law and the rapid pace of improvement in computation speed and network design, it is necessary to transition most wireless communications management to equipment-based sharing protocols and away from bandwidth-management schemes. This is beginning to happen, as we see the phenomenal success of wireless LANs of various forms. We see its emergence in the FCC's approval of the U-NII Band, the UWB order, and the software defined radio order. As of this writing, the FCC is showing signs of adopting even broader spectrum commons oriented policies. If wireless networks indeed become more accessible and less licensee-controlled, then a business focus on building smart communications devices and smart services in the generally-accessible network, rather than models built on controlling the eyeball of the consumer by controlling the means for reaching it are, in the long term, more socially desirable and, quite possibly, more agile and sustainable in the face of rapidly changing technological conditions that are destabilizing the owned-spectrum model.

It is also the case that focusing on replicating the mass media model and bringing it to the 21st-century is a mistake. The mass media model is a result of the constraints of a particular capital structure of the delivery devices capable of transmitting information to large populations over a distance. From the introduction of mechanical presses more than 150 years ago, through the introduction of movies, phonorecords, radio, television, cable systems, and satellites, the capital cost of communicating effectively has driven toward centralized, commercialized production, with stark separation between production and consumption, and large returns flowing to a small number of professional "speakers" or "performers." Yet even in this period, people spent more on talking to each other—over the phone—than on listening to or viewing others. The dramatic decline in the cost of capital necessary to communicate effectively to large audiences over a distance, represented by the Internet, provides an opportunity to reverse the relative emphasis. It is not that mass media professionally produced commercial content will not be important. But it will be secondary to tools enabling individuals to band together in communities of interest, to communicate with each other and be users and storytellers, rather than passive consumers. The success of massive multiplayer online games like Everquest or Ultima Online; of sites like eBay; of practices like Napster, or the success of SMS, is indicative. The focus on replicating the mass media model in this environment is a mistake because that is not what people value most. It is simply what the relatively crude end-user equipment available in the past made most deliverable. The result was culturally important industries that are comparatively small to the industries devoted to giving people tools to communicate themselves.

To get the full extent of how much more people value the ability tools with which to communicate with each other than the ability passively to receive from others, all one need do is look at the 2001 Statistical Abstract of the United States.[1] The entire sound recording industry, one of the two primary driving forces behind the expansion of copyright, had receipts of roughly 12.25 billion-dollars in 1999.[2] All movie and video (53)[3], all radio and television (47.6)[4], all cable (60), put together with music, accounted for 170 billion dollars in receipts for 1999. By comparison, receipts from telecommunications services (319),[5] data and online information processing (57),[6] combined with only the domestic income of[7] software (214), computer hardware (226), and communications equipment (49) yields revenues of 865 billion, more than five times as much. In other words, the industries involved in making it possible for users to make, store, and manipulate information dwarf those dedicated to vending finished information to passive consumers.

The enormous value that users place on tools for communicating, generating, and manipulating information and cultural artifacts suggests a long-term conflict between two business models with two fundamentally opposed structures of where value is, how it is maximized, and how it is appropriated.

Business models epitomized by the recording industry, the movie industry, and traditional television broadcast, focus on attracting large numbers o people who will passively receive a stream of information or cultural goods, and consume them as and under the conditions set by the vendor. The value in that model is in the content, and specifically, in its production values. It is maximized by heavy investment in demand management and taste shaping, intended to get large audiences to fit their tastes to the output of the producer. The star and celebrity system is the central operational aspect of this strategy. Value is appropriated primarily in one of two approaches, which in some mix or another are employed by all these industries. The first is sale of the cultural products as packaged goods, with control over use of the goods so as to facilitate price discrimination. The sale as packaged goods approach is most clearly articulated in the recording industry. The price discrimination aspect is most clearly articulated in the movie release window approach. As to both aspects of this approach, control over uses that users make of the good is seen as an absolute necessity. The second approach is to broker the attention of consumers to advertisers. Television is the most obvious case of this. But obviously movies and videos all also include a component of attention brokerage, from the commercials before the movie to the product placements in them. Again, the vendors need to control how users use the goods, because they must secure their attention. The gross example is the one quoted from Jamie Kellner, head of Turner Broadcasting, claiming in an interview with Cable World in May 2002 that when you the consumer fail to watch the ads, "you're actually stealing the programming." As just noted, the total revenue of this business model in 1997 was about 170 billion dollars.

The competing business model is focused on allowing users to use the network, rather than consume finished information goods delivered over it. The computer hardware and software industries, and telecommunications services and equipment industries epitomize it. The value in these models is generated at the end points of the system. It is represented by the value to consumers of all the behaviors and interactions that the tools provided to them enable them to do over the life of the equipment of service. It is maximized by making the tools either highly versatile to allow users to re-purpose them as their needs and demand changes over time, or by optimizing the tool for a particular kind of valuable activity. It is appropriated by the provider in the price consumers pay for the tool, be it an up front payment for the equipment, which builds in the future value of use into the valuation of the equipment, or in service charges paid over time. As noted earlier, the value generated through providing users the tools they need to manipulate information and culture and communicate with each other was over five times the value generated by the other business model.

It is important to understand that these business models are not complementary, but are instead ecologically competitive. The first model needs to constrain the set of moves consumers can make, in order to make sure that they buy their packaged goods and give them their attention. The second model needs to allow users as much freedom as possible to create, get, manipulate, and send information and culture, because this is what maximizes the users' value over time. Crisp examples of this conflict are the litigation over the introduction of the VCR, and more recently the various devices that give viewers more control over television, like TiVo or SonicBlue Replay. Apple Computers' "Rip, Mix, Burn, It's Your Music" ad was a particularly bald statement of this conflict. But the steady stream of bigger hard drives, CD burners, soundcards and speakers from the desktop and laptop manufacturers, and the improving mp3 players from mainstream companies like Microsoft and Real Networks, speak louder than any words could.

The question, seeing these facts, is why on earth anyone would want to think of mobile wireless devices as "mass media" extensions? Mobile wireless providers should be working to develop stronger business models for putting tools in the hands of users that will be most valuable to those users as tools for communication and information processing. Doing so will place mobile wireless squarely in the camp that is faster growing and higher value—the camp of the toolmakers. To adopt an approach that tries to make the mobile wireless device into a glorified car radio, on the other hand, is to choose the lower value trajectory.

3 The Policy Battle

I have suggested up to this point that exclusive rights in information are likely to impose a levy on service providers that try to migrate their wireless systems to a mass media model. The "mass media" approach to mobile wireless, however, should be abandoned. Mobile wireless should focus on business models oriented toward providing tools for users to make and exchange their own information and cultural expressions. It is on this model of toolmakers that companies have provided the most value to users. They have been able to appropriate it through sales of equipment and services that enable users to make and communicate their own information and culture, rather than consume ready-made culture like so much packaged cereal.

Exclusive rights are, however, also likely to pose substantial limits on what either equipment manufacturers or service providers who wish to offer high-speed access can do. Whether that will be the case, and to what extent, is still open for political battle, a battle that toolmakers have been slow to join. Because law and regulation are central pillars that enable the copyright indus tries' model to survive, and because of their cultural salience, the copyright industries have a tremendous power base in Congress. They have succeeded in passing quite extreme measures, like the DMCA, and quite blatant wealth transfers, like the Sonny Bono Copyright Term Extension Act of 1998, that provided a retroactive windfall extension on copyright inventories. Quite magically, the passage of these laws is rarely perceived as a partisan issue. Toolmakers, on the other hand, have been less dependent on regulation, and have been present only in telecommunications vis-à-vis sector-specific regulation, but not usually in the politics of exclusive rights. The carrier exemption from ISP liability in the DMCA is a relatively rare example where the toolmakers have been involved in, with some success, the exclusive rights legislation process.

Given the long term conflict between the toolmakers and the culture-as-goods vendors and attention brokers, the failure of the toolmakers to be present when the institutional and legal terms of this battle are being set in Congress is a strategic error. Toolmakers—computer hardware and software manufacturers and service providers, and telecommunications carriers and equipment manufacturers—must understand that the institutional ecology that sets the parameters for how valuable their tools and services could be is being set in the exclusive rights legislation arena. These businesses have a long-term interest in securing as robust a system of free information and cultural use and exchange. That is the way to maximize the value of machines and communications to users. And it is that value that users end up being willing to pay for to those who build the tools and services that make them possible.

Endnotes

1 U.S. Census Bureau, Statistical Abstract of the United States, Section 24, Information and Communications (2001).
2 Table No. 1120, NAICS 5122.
3 Table No. 1120, NAICS 5121
4 Table No. 1120, NAICS 5131.
5 Table No. 1120, NAICS 5133.
6 Table No. 1120, NAICS 51514191, and 5142.
7 Following numbers all from Table No. 1122.

References

Benkler, Y. (1998). Overcoming agoraphobia: Building the commons of the digitally networked environment. *11 Harv. J. L. & Tech 287.*

– (2001). Siren songs and Amish children: Autonomy, information, and law, *76 N.Y.U. L. Rev. 23.*

– (2002). Some economics of wireless communication. *15 Harv. J. L & Tech.*

– (2004). Peer production of survivable critical infrastructures. Presented at TPRC 2004, retrieved from the World Wide Web: http://web.si.umich.edu/tprc/papers/2004/340/Benkler%20Critical%20Infrastrcutures.pdf.

Litman, J. (1987). Copyright, compromise, and legislative history. *72 Cornell L. Rev. 857.*

Noam, E. M. (1998). Spectrum auctions: yesterday's heresy, today's orthodoxy, tomorrow's anachronism. Taking the next step to open spectrum access. *In Journal of Law and Economics 56*(2), 765-790.

United States Census Bureau. (2001). Statistical abstract of the United States.

Werbach, K. (2001, November). Open spectrum: The paradise of the commons. *Release 1.0.*

15
3G or Not 3G: The WiFi Walled Garden*

*Kenneth R. Carter***

1 Introduction

The term "wireless network", as used in common parlance, is much of a misnomer. The word network actually means a system of interconnected relationships. This system derives its functionality from its ability to provide switched connections. However, the term wireless refers to the network's ability to provide mobile connections, though wireless technologies only represent a small portion of the network's elements. The elements of a wireless network include not only the wireless connection, but also switches and landlines. In most wireless networks, the wireless connection only provides the so-called "last mile" connection to the end users and the connection provided by radio is usually not switched at all. The wireless part of the network is the mere tip of the iceberg.

This distinction is more than mere semantics. It is an important fact to remember when analyzing the economics of wireless networks. The fundamental challenge for network operators as they enter the third generation of cellular networks is obtaining strategic control of all parts of the network. There are two entities on either side of its value chain which it must actively cultivate. On the one side are the network's subscribers; on the other, content. To attract subscribers, the network must provide appealing content, and vice versa. A network must reach a critical mass, to sustain itself, otherwise it will not be economically viable. Without the ability to control network access, the carrier lacks the ability to cultivate these two entities.

Carriers need to translate content into services that are not only appealing for customers but also that contribute in making their lives easier and increase the company's revenues. Wireless terminals are moving beyond fancy cordless

---

* I would like to acknowledge the helpful comments received from Prof. A. Michael Noll in preparing this chapter.

** With assistance from Valentina van der Dys.

phones, becoming mobile computers, walkmans, and TV sets. Strategy and investment will be reliant on what proved out to temporary and what are permanent changes in technologies, applications, and adoption. Consequently, the deployment of next generation wireless networks has languished.

Further compounding this dilemma is that one of the barriers to entry — the exclusive lock on the spectrum connecting the end user — is falling. The system of spectrum allocation whereby carriers purchase the exclusive right to use spectrum has served as a barrier to entry for new firms. Recently, networks which employ unlicensed spectrum, for which the carrier has not paid for the right to use the frequency, have begun to be used to provide mobile data communications. These "unlicensed" networks are perceived as a threat to the profitability and commercial viability of existing cellular networks and emerging 3G networks. Next generation service providers will have to integrate licensed and unlicensed spectrum in their networks and find means of recreating the barriers to entry of licensing regime with differentiation, externalities, and network investment.

2 The U.S. Wireless Industry

In the United States, wireless communications is very heavily regulated. Nearly every emission of electromagnetic waves is subject to prescription by U.S. statute or Federal regulation. Since the late 1920's, the need to regulate the broadcast of radio signals into the ether has became apparent due to the fact that radio signals cause interference, and at a certain level of interference no one can clearly receive signals. The Federal Communications Commission (FCC) has historically assigned bands of adjacent frequencies to particular applications, then allocates the exclusive right to those frequencies to minimize the problem of interference. The FCC has held various auctions to ensure the economic allocation of this scare resource. Potential wireless providers do not actually bid for spectrum, but rather a license granted by the government for the right to emit electromagnetic waves, into the ether at a given frequency power lever in a specified geographic location. The recipients of licenses in these auctions must then make further capital investments in network infrastructure to provide these services.

There are, however, certain bands for public use. This spectrum does not require a license for use, but the use must conform to FCC rules. These low emission devices, such as remote controllers, wireless LANs, cordless phones, and garage door openers are governed by Part 15 of the FCC's rules. Moreover, the FCC rules previously prohibited transceivers from operating on multiple frequency bands. Despite the fact that CB radio and marine VHF radio might be similar equipment, the transceivers were required to be in separate boxes. In September of 2001, the FCC changed its rules on Software Defined Radios (SDR), which might now allow radios to operate on multiple standards and services. SDR employs computer processing to enable a single transceiver to provide multiple modes, technologies, platforms, and protocols.

This policy shift reshuffles the spectrum deck for service providers, networks, and equipment suppliers (Bauer et al., 2004). It permits third-party applications and access to other network protocols offered by other types of providers, provided that it conforms to the FCC's software defined radio rules. These three items: Licensed spectrum, unlicensed spectrum, and SDR can be combined to afford a powerful tool for carriers to deploy of advanced, spectrum-hungry 3G services.

3 The Move to 3G

As new wireless products evolve, this market is facing discontinuous change, making it impossible to predict the future. Predictions of consumer demand are virtually useless; a scenarios analysis may prove more reliable and more useful.

3.1 Technological Evolution — Next Generation Networks

Modern mobile wireless networks can trace their origins back to technologies developed at Bell Labs in the late 1940s. Cellular networks are designed to localize the wireless connection and reuse those frequencies in other parts of the network. Cellular networks derive their name from the system of localized, low power base stations that cover a specific area. The base stations are sited to give overlapping coverage, fitting together like cells in a tissue. A set of channels are assigned to each cell. The channels, because they are low power, can then be reutilized in adjacent clusters. This is as opposed to having one centralized, powerful antenna broadcasting over a large area.[1] However, modern cellular networks were not deployed commercially until 1984. This is due to the fact that inexpensive computers were needed to handle the switching and "hand-offs" need by the network.

The original cellular networks of the 1980s, referred to as first-generation (1G) analog networks, were only capable of providing voice communications because they employed analog technologies for the wireless link. In the mid-

1990s, the FCC licensed the PCS bands (Personal Communications Service), which is referred second-generation (2G). It has dedicated channels for both voice channels and optimizes voice traffic. This technology maintains a dedicated channel as long as the call is maintained, regardless of the use. However, PCS uses digital modulated spectrum to provide not only voice, but also limited data communications such as text messaging, email, voicemail, short message service (SMS), and caller ID. Major U.S. carriers use CDMA, TDMA, GSM and PDC to provide 2G networks. 2G users get low transmission rates, usually lower that 9.6 Kb/second per time slot. If new services will focus on data transmission, with this rate of transmission and inefficiency, it will not be possible. Some cell phones are capable of providing limited text browsing of the Internet. This is often referred to as 2.5G.

Cellular networks are now entering what is known as the third-generation or "3G" platform. There is no standard definition of what 3G is, but is generally accepted to mean a wireless network capable of providing high-speed data connectivity which is comparable to current fixed-line communications. 3G is touted as being able to offer broadband services, packet based transmission of text, digitalized voice, video, and multimedia. 3G networks include the capability to support circuit and packet data at high bit rates:

- 144 kilobits/second or higher in high mobility (vehicular) traffic·
- 384 kilobits/second for pedestrian traffic·
- 2 Megabits/second or higher for indoor traffic

Other services and capabilities include:

- Fixed and variable rate bit traffic
- Bandwidth on demand
- Asymmetric data rates in the forward and reverse links
- Multimedia store and forward.

The uptake of 3G has been relatively slow in the U.S. The technology provided by 2.5G is already allowing carriers to deploy a wide variety of features in their wireless network. But, its major limitation is the speed of data transmission. Despite the fact that wireless connection speeds are greater than the wireline speeds of the past, the connection speed of 2.5G networks may be enough to provide voice grade service as well as multimedia content to satisfy customers, obviating their demand for more advanced networks.

Since the wireless network is emerging as an all-purpose, mobile communications network, estimations on total market size will be highly relevant Next generation technology will provide improved services to the market with higher speed connection and increased broadband. Pervasive competition could arise if the vast majority of carriers act aggressively in providing new and targeted services to their customers. Marketing and segmentation tools are extremely important in this strategy. They could deploy a differentiating strategy that consists of targeting a specific market niche. In this practice they will be skimming revenues in the segment of those subscribers that demand data transmission on the go. They should concentrate in launching as many services as they can, collecting the advantages of both skimming practices and being the first in the market.

One emerging trend is carriers offering WiFi instead of traditional cellular. Many service providers offer "franchises" of their network, becoming wireless wholesaler, offering service in chain stores, airports, and other interested companies. However, the lack of barriers to obtaining this unlicensed spectrum and the relative low cost of setting up WiFi hotspots allows customers to use their computers on a variety of networks. This undermines the service provider's ability to institute consumer loyalty, reducing churn and increasing ARPU. One approach is when carriers act as Internet portals to their customers. For example, web portals like Yahoo! have been offering other wireless email option by allowing their customers to access their web site using cellular phones. By doing this, carriers control their customers' Internet access and navigation path.

3.2 Economics

While one may not be able to predict the demand for 3G, its services, or its incarnation, with a basic understanding of network economics, we can start to understand how new networks and wireless applications are likely to be used. Despite the advent of pervasive new technologies, the general rule of wireless will continue to be: plug it in unless it has to move. There are some notable exceptions to this rule.

Even assuming the fixed cost of the radio equipment is similar to the equipment cost of a wireline connection, the increase function of modulating and demodulating radio signals adds delay compared to when those signals are conducted over a conduit. In economic terms, the radio connection lengthens the production process in transmitting and receiving signal, and therefore the average cost of transmission.

In this trade-off, wireless connections should be used when the advantages of radio communication outweighs its additional cost. Such advantages are when the application must be mobile, portable, or wiring is overly cumbersome. An examples of an overly costly wiring installation is when the installation of conduits requires channeling through walls or digging up streets. A wireless solution such as Bluetooth computer peripheral devices might be used where the wires are overly cumbersome where six USB port connections can be replaced with a single wireless hub. This is similar to ad hoc networks such as a ham radio, which could not be easily formed when a new wire must be used to connect an incremental user. Radio connections also allow for greater shared costs when there is complementary, non-rival use. Sharing of the spectrum is easier than sharing lines, especially for last mile connections.

This wireless-wireline optimization also appears in building a wireless network. The architect of a network has to balance two competing costs: spectrum and network hardware, network architecture balancing cell size and efficient use of spectrum versus the cost of hardware (hubs, routers, etc.), and the cost of wired networking. On the one hand there is the cost of the size of access point and on the other, cost to wire it up. The cost of spectrum places limitations on the size of the cell site. A network provider could use unlicensed spectrum, such as those permitted under Part 15 of the FCC's rules. However, these applications are required to use significantly lower power which implies a much shorter range. In sum, if the network uses unlicensed spectrum, while it does not have to pay for those licenses, it must spend significantly more to wire up much smaller cell sites. Using these tradeoffs, the network engineer can model the cost of spectrum versus the cost of wireline network and optimize that expenditure.

3.3 Network Externalities and Game Theory

An externality is any economic effect that is felt by a third party not part of the original economic transaction. A positive network externality is usually the increase in utility of the network created by the non-rival addition of new subscribers. As the number of users increases, the value of the network to all participants increases exponentially. This is known as Metcalfe Law. A network must reach a critical mass, to sustain itself, otherwise it will not be economically viable. Beyond the critical mass point, the network experiences natural growth as each new user decreases average cost and increases the effect of network externalities.

The effects of network externalities are most apparent in the interconnection of networks. The utility of a network is directly proportional to the number of users. A larger network would not interconnect with a smaller one because the marginal benefit to the smaller network's users would be greater than the benefit to its customers. Networks try to grow their size and exercise market power to tip the network in its favor. Since the economics of network externalities, tipping, and rules of game economics are relatively well understood by the industry participants, there are a limited number of strategic options and responses. This makes game theory highly relevant and carriers must carefully watch and anticipate the moves and responses of its competitors.

To take advantage of network's externalities, the number of customers should be optimized. To carriers, size and network design is extremely important. This presents a chicken and egg dilemma for the network. To attract subscribers, the network must provide appealing content, and vice versa. A network must reach a critical mass, to sustain itself, otherwise it will not be economically viable. This is most evident in the wide geographic footprint of the network necessary to provide a wide roaming area for mobile users.

4 The WiFi Challenge: 3G and Unlicensed Networks

4.1 The "Free" Spectrum Challenge

The system of licensing whereby carriers obtain the right to use spectrum through auctions has the unintended consequence of serving as a barrier to entry for new firms. Conversely, networks which employ unlicensed spectrum, for which the carrier has not paid for the right to use the frequency, present a threat to the profitability and commercial viability of existing cellular networks and emerging 3G networks. New service providers are beginning to offer portable Internet access for laptops and handheld computers in airports, hotels, cafes, and other public places. Five different hotspot strategies have been identified:

1. Individuals or companies who install in commercial places
2. Aggregators who combine local installations to provide a national foot print
3. Major wireless service provider offerings
4. Computer and electronic manufacturer consortia
5. Grass roots individuals offering free or low-cost access

Most ventures do not rely on a single mode of entry but are pursuing a combination of these strategies.

When a network operator chooses to install hotspots in partnership with another commercial entity, the offering takes advantage of the special expertise derived from each provider in the partnership. One of the early movers in this arena is T-Mobile, a wireless service provider. T-Mobile made headlines when it purchased a company with contracts to place wireless hotspots in Starbucks coffee shops. Starbucks is offering three subscription plans: a $29.99 per month unlimited plan with a 12 month commitment; a month to month unlimited plan for $39.99; and metered plan for $0.10 per minute with a 60 minute per connection minimum. An organization like Starbucks, clearly not a network operator, finds it more cost effective to outsource Internet access to an organization that specializes in providing network services. Starbucks anticipates that having the Internet access available for its customers will help sell a greater number of $3 cups of coffee. As a PCS operator, T-Mobile can take advantage of its existing mobile service infrastructure to leverage the build-out of more geographically dispersed WiFi services. While T-Mobile does not actually sell access to unlicensed spectrum, as it does with its cellular service, it can offer connectivity to the Internet on a subscription basis using unlicensed spectrum.

An alternative model is the complimentary offering model which McDonald's Restaurants has decided to pursue in 2003. McDonald's Restaurants announced it has selected Cometa Networks to provide WiFi service as it begins to test market wireless Internet service in three U.S. cities. McDonald's recently began offering one hour of free WiFi access to anyone who buys a combination meal in one of ten stores in Manhattan. The company claims that it will extend the service to 300 stores in New York, Chicago, and another city in California. Cometa subsequently filed for bankruptcy protection.

Potentially worse for the prospective 3G network provider, there may be a complete end-run around the commercial wireless provider. A popular activity among computer hobbyists is "warsniffing", traveling around with the goal of gaining "free" Internet access using a legitimate, but unprotected, W-LAN connection. "War-chalking", taking the information learned about open W-LAN connection and creating a map, sometimes leads to the publication of these maps in so-called "weblogs".[2]

It is feared that these "free" spectrum networks could present a threat by cannibalizing their existing businesses. The carriers have paid handsomely for their licenses, and the upstart WiFi carriers do not face that cost structure. The license is generally considered to be a sunk cost and does not effect pricing necessarily competitors do not have that cost. Moreover, since WiFi networking is an open system of protocols the carrier looses the control of attachment of handsets and terminal equipment to its network. This lowers switching costs and lessens ability to charge a premium.

However, this fear may not be as threatening as initially perceived. Any system based on "free" pricing is doomed to fail once scarcity, or rival uses for the finite good, is introduced. The owner of a private hotspot, such as campus

W-LAN or home WiFi access point, will institute access protection as soon as the use of "war-free-riders" start to negatively affect his use. For example, if a "war-sniffing" neighbor has gained access to your home WiFi and is using your cable model or DSL while you are at work, you are not likely to care very much. However, when you return home, you would be unlikely to share your bandwidth with the interloper. Until recently, one of the sharpest criticisms leveled at WiFi is the lack sufficient levels of encryption to prevent the eavesdropping on data and that each employs only rudimentary means to block access by would-be hackers. Implementing security features adds support and configuration costs for both end users and product developers. Nonetheless security features are being incorporated in new products. Furthermore, the WiFi operator cannot guarantee a level of network performance because WiFi is not granted any interference protection under the FCC's rules. Thus, "free" hotspots are likely to disappear relatively quickly.

Moreover, the McDonald's-style model of complementary WiFi is proliferating and may present a serious competitive threat to cellular carriers' efforts to enter this market. However, complementary WiFi may, in fact, prove antithetical to McDonald's fast food business. McDonald's service operations are engineered to get customers in and out of the door; the more and faster, the better. McDonald's stores are also designed with hard plastic seats and other fixtures aimed at getting the customer out of the door in less than 20 minutes. And while most of the McDonald's-going-public is unlikely to bring a laptop to the drive-thru, those who frequent cyber cafés might. A store like Starbucks is very different from McDonald's in that it wants the customer to linger and make repeat purchases.

What is needed is a carrier who can integrate these platforms.

5 Building a WiFi-Proof "Walled Garden"

As we have seen, spectrum-based barriers to entry are insufficient in and of themselves to provide carriers with a sustainable competitive advantage. Differentiation is the name of the game and in this game, free spectrum may in fact help carriers. It creates a new, albeit undifferentiated, product to add to a suite of wireless products. Granted, a network carrier may see some cannibalization of its business from the alternative mode, but these are some non-exclusive, presumably lower quality services. The good news for the carrier is that it now has the ability to varying grades of service quality at corresponding price points. This affords the carrier the ability to price discriminate among its customers. If done right, price discrimination can increase profits significantly. The dilemma the carrier faces is how to prevent potential customers from making an end-run to other "free" networks.

In creating a walled garden, network providers control access to content, limiting the availability to that which they benefit from the transaction. The openness of a carrier's network ranges from totally proprietary to contract carriage to common carriage. There is an optimal height and number of gates walled garden so as to provide incentives for others come and plant their tulips in it.

5.1 Differentiation and Price Discrimination

To compete effectively, these providers must find ways to differentiate their products. By allowing just any device to attach to its network, a provider can attract more users, but simultaneously runs the risk of turning wireless Internet into a free-for-all. Unlike the cellular network paradigm in which only approved phones are allowed to connect to a network, WiFi service providers currently have far less control over the terminal equipment which can connect to their networks. In an open, competitive environment, there are virtually no impediments to the user switching to another provider since his device can also be used on other (presumably competing or free) networks. To attempt to differentiate themselves, carriers may find some means of offering terminal equipment that is not completely interoperable with the networks and features of other carriers, or at the very least, equipment designed to attach to its primary provider's network first (Noam, 2002). This differentiation will afford the opportunity to price discriminate.

In general, price discrimination improves a firm's financial performance by extracting surplus consumer welfare. However, problems arise in price discrimination. These include when the products are so cheap that a difference in price is hardly noticeable to the consumer, the pricing does not reflect products underlying cost structure, or there is little difference between the competitive products which would afford the ability to discriminate. Another impediment to price discrimination is when the provider cannot differentiate users such as those in a peer to peer network without connection through the network provider's facilities, including both the spectrum and wired portions of the network.

It is possible to make a business selling what is normally a free good. Let me explain in this way. People buy air which has been dehumidified, filtered, and compressed into tanks for SCUBA diving. They will pay for a commodity which has be subject to some sort of differentiation which makes it more useful. The SCUBA diver is paying for the pressurization and not the air. Similarly, most people routinely buy bottled water, paying for the convenience of refrigerated water in a container and the perception of purity. This has become a very profitable business since the mark-up on water could be greater than that of Coca-Cola, which requires ingredients other than just water.[3]

Using its exclusive as well as nonexclusive spectrum, a wireless carrier can now discriminate with high and low quality products. For example, cell phones which incorporate SDRs which allow them to be used as CB radios[4] or walkie-talkies. This might not be a practical application for business use because of the risk of interference or interception, but it would probably be quite acceptable to more casual users. This is a one-part pricing scheme whereby a cell phone user does not pay to use his phone outside of the initial purchase. This means that once the device is purchased, the consumer does not pay for continued use of the device and there is not necessarily a continued relationship with the vendor or network provider. However, the carrier wants to collect airtime and other service fees. Ultimately, a phone using unlicensed spectrum seem likely to be cheaper to the end user, or at least he avoids reoccurring charges. Carriers can free up the utilization of their licensed bands to provide high quality service while letting other services "ride steerage" with the "unwashed bits" using the unlicensed bands.

5.2 Content and Conduit

To date, cellular carriers in the U.S. have not taken an active interest in pursuing 3G strategies. Cellular offerings are still traditional telecommunications in new packages without wires. Each company has its own proposition to the market, mixing the basic component such as coverage, tariff plans and features. The use of the network is for voice and data transmission of cellular customers. Much of the value created in next generation services will be in content creation and distribution. Many carriers, the progeny of the Bell System, lack understanding of marketing media products which is idiosyncratic. Telecommunications companies have traditionally failed at media offerings and are not likely to reinvent themselves as something they are not. Rather, these carriers should focus on their existing networks, complimenting them with WiFi hotpots. Their principle asset is the local customer. Thus, existing cellular carriers already have advantage of existing network they can leverage to build out and offer WiFi services.

Cable companies are more likely to benefit from the deployment of WiFi hotspots. Growth in the sales of WiFi gear will have positive downstream implications on the demand for complimentary products and services such as high-speed Internet access. WiFi coupled with broadband is generating a procyclical adoption pattern. Both cable and DSL modems are being sold already equipped for WiFi. Since cable modems have proved to be more successful than DSL, it is likely that cable companies will capture more of the benefit from this increased demand.[5] Moreover, cable companies, often setup more like media firms than network providers, will have an opportunity to distribute news and entertainment offerings through wireless channels. These would be through bundling and cross-marketing of complementary products, which they currently offer to residential subscribers. Cable companies further have the resources to sell advertising or sponsorships used to support the deployment of next generation services. Nonetheless, cable companies face the challenge of creating a national footprint which is an essential asset in offering wireless services.

A few scattered WiFi hotspots alone are not sufficient to create a viable wireless network. A nation-wide, ubiquitous network is necessary. Each hotspot would be useless alone if it were not connected to other networks. From there the connection is made to other cell sites, local and long distance telephone networks, or even the Internet. A wireless network also requires a centralized database in order to keep track of where an individual user is, so that an inbound call can be routed to the cell cite serving the user. Presumably, most users of cellular networks are mobile. When a mobile user travels from one cell site to another, the system provisions service until the customer physically gets under the coverage area of the new cell site.

6 Conclusion

Unlicensed spectrum, such as WiFi lowers the barriers to entry to the market. However not completely open. Those carriers which implement strategies viewing WiFi as a complete threat or a complete means of entry are likely to fail. A system based on the sale of hardware, without services or differentiation is likely to result in a market of commodity products. As a business strategy, it is ultimately indefensible and hence not sustainable against competitive entry. WiFi will be an intrigue part of any network wireless or business strategy. Carriers must have a balanced offering of services within and outside of their networks. In the final analysis, carriers will continue to organize cartels and rely on other barriers such as scale and sunk costs to exclude new entrants. In the world today, oligopoly is a prevalent form of market structure because of its stability. In an oligopolistic market, the offerings oftentimes are not differentiated. In our case, spectrum is spectrum. What would make an oligopoly is if only a few firms had control or possession of most of all of the available spectrum licenses. As with any oligopoly, a barrier to entry must exist. With a spectrum license, the barrier to entry is that the government can auction only limited amounts. This is a natural barrier to entry, because it is basic to the structure of the telecommunications market in its current state. Furthermore, because the government regulates the availability of spectrum licenses, an incumbent may not even need to initiate strategic actions to deter entry.

Managing a firm in an oligopolistic market structure is complicated, because all decisions, especially pricing and investment decisions, involve important strategic considerations. Because only a few firms are competing, each firm must carefully consider how its actions will affect its rivals, and how its rivals are likely to react. The strategic considerations can be complex. Furthermore, decisions, reactions, reactions to reactions, and so forth are dynamic, evolving over time. When managers evaluate the potential consequences of their decisions, they must assume that their competitors are as rational and intelligent as they are. Then, they must put themselves in their competitor's place and consider how they would react.

WiFi is merely one end user link in a larger, integrated network.

Endnotes

1. The low power has the added advantage of reducing power consumption and potential health risks.
2. This activity has been dubbed "war-sniffing" after the 1983 film War Games. In the movie, Matthew Broderick breaks into a NORAD computer by randomly dialing into computer modems. War-sniffing comes in several different flavors; "war-walking", "war-driving", and even "war-flying", depending on the kind of vehicle one uses. "War-spamming" is the use of an unsecured access point to send spam email on the Internet, and "war-jacking" is a denial-of-service attack that knocks a one hotspot in favor of the hacker's. The FBI has demonstrated a keen interest in many these practices. Since WiFi devices are afforded no interference protection under Part 15 of the FCC's rules, war-sniffing may not be illegal per se, depending on what the would-be hacker does once he has accessed the unprotected access point.
3. Thanks to Robert Pepper for the illustration.
4. Technically, CB radios are licensed by rule and are not unlicensed. For the purpose of this illustration, the distinction is moot, because a CB operator does not have exclusive access to the spectrum.
5. WiFi is a double-edged sword for both cable and telecommunications companies offering broadband products. While it stimulates demand for broadband access, it can also be used to provide an Ethernet for users located in the same building or complex. An Ethernet using WiFi is generally less costly to set up rewiring the

building. The sharing of a single broadband access among these users may reduce the demand for connectivity to the building.

References

Bauer, J., Lin, Y., Maitland, C., & Tarnacha, A. (2004). Transition paths to next generation wireless services. Retrieved from the World Wide Web: http://faculty.ist.psu.edu/maitland/transition%20paths.pdf.

CTIA. (2004). *Semi-annual wireless industry survey.*

Noam, E. (2002). Opening the Walled Airwave (pp. 35-55). In R. Entman (Ed.), *Telecommunications competition in a consolidating marketplace,* Aspen, CO: The Aspen Institute.

Telecommunications Industry Association (TIA). (April 10, 2002). *Telecommunications Market Review and Forecast.*

16
Emergency Communication Needs: Mobile Content

Jonathan Liebenau

1 Introduction

The attack on September 11, 2001 brought a new awareness of the utility of mobile communications in times of emergency. It also brought to attention two things that had been largely overlooked in the rush to promote the use of mobile phones: the frailty of wireless networks and the limited assistance that they can offer in emergencies. Most of the subsequent discussion since has focused on questions about the reliability of the systems in their ability to work when congested and when relay stations have been damaged or destroyed. However, neither before nor to a significant degree since has there been much focus on the other central problem: How can the content of communications over mobile networks contribute to solving problems in emergencies?

In this chapter we ask two kinds of questions: First, what content would users want from mobile phones and how might they use them? Second, who would provide such content and what might be the means by which it is managed and maintained? First let us review what current emergency uses are and how people use that content and functionality.

2 Emergency Communication Needs

Mobile communications worked to some degree well on 9/11 in the World Trade Center and in the hijacked airplanes. However, the mobile system became predictably overloaded in the regions affected, with domino effects nationwide. Small data transmitters such as Blackberrys and pagers worked as limited alternatives (Kapsales, 2004). Nevertheless, they demonstrated the imaginative ways people sought out means to communicate, and showed that parallel technical systems can work in emergencies.

This disaster has sparked interest in the significance of certain kinds of emergency communications content that must now be regarded as high priority. This includes what was formalized following the Kobe earthquake in Japan as the "I am alive" (or IAA communication) function. This occurred in the form of widely reported personal communications from within the World Trade Center, in the form of ad hoc web sites established immediately following the destruction of the buildings, and in hundreds of human-interest stories in the press (Noam & Sato, 1995; Noam, 2001). We all saw the importance of providing anyone the opportunity to contact loved ones when caught in an emergency. Many such people sought advice, some of which might have been translated into tangible help such as instructions on how to escape danger. The deadly tsunami that struck the shores of the Indian Ocean in December, 2004 is another example.

Except in times of disaster we usually regard communication systems primarily as requirements for our social and economic needs, with special functions for national security and other governmental activities, plus emergency services such as police and firefighters (Anderson & Gow, 2000). When we suffer a disaster, however, we are starkly reminded of the utility of systems for saving lives, directing recovery work, and performing other highly valued functions, such as communicating with loved ones in extreme conditions. On September 11 we learned of the relief felt by those able to make contact with family from the streets and of friends and colleagues able to find each other from within the chaos of the escaping crowds. However, we also learned of the frustrations of those who encountered broken or busy lines, of data lost, and priority users unable to use dedicated systems.

In subsequent years companies and goverment bodies have focused on the considerable needs, and capabilities, in ensuring greater resilience and access. The possibility of increased interoperability is of particular interest, especially since the inhibitions so far have been more in the realm of regulatory practices (the allocation of spectrum) and competition policy. Military applications of equipment capable of spectrum switching have long existed, but for legal as well as commercial reasons they have until recently not been built into civil systems.

One of the surprising elements is that although questions about the reliability of wireline networks have long been a concern, the only special consideration that wireless networks have attracted in the literature has been the availability of lines, and the establishment and roll-out of enhanced emergency calls.

There have been several rulings by the Federal Communications Commission concerning the criteria for enabling 911 calls to be made from mobile telephones (FCC, 2001). The basic requirement, as specified in a series of orders since 1996, has been to improve the quality and reliability of 911 emergency services for wireless phone users, by strengthening rules to govern the availability of basic 911 services and the implementation of enhanced 911 for wireless services. Two technical goals are at the heart of these enhancements. One is the ability of any wireless telephone to override subscription status and complete any 911 call. The other is to enhance the ability of emergency services to find where an emergency call originated from by means of location identification data. These requirements have been phased in since 1999, though incompletely.

We can get some idea of the significance of this functionality by considering how people behave in emergencies when they have mobile phones (IBM, 2002). The fact that mobile phones are ostensibly commonly acquired for emergency use is revealing of people's expectations. Some studies of the extent to which the phones are used in emergencies show that although most people do not have occasion to make such use, the occurrence is common enough to regard the likelihood as reasonable. There are distinct differences internationally with regard to emergency uses and until recently the high level of ownership but often low level of usage in the U.S. was partly explicable by people acquiring phones solely for emergency use (Palen et al., 2000).

Chapman & Schofield (1998) determined through research in Australia that 1 in 4 users have reported a dangerous situation; 1 in 8 a traffic accident; 1 in 16 a non-road medical emergency; 1 in 20 a crime; and 1 in 45 being lost in the bush or being in difficulty at sea. There are also uses for more direct medical applications, such as transmitting ECG results to hospitals so that commencement of appropriate therapy can be organized more rapidly upon arrival. Other work on the sociological and psychological aspects of the use of mobile phone indicate that people in any case come to depend on connectedness (Geser, 2002) and the implication is that people, women in particular, use the mobile to ward off discomfort and perhaps fear even in normal circumstances (Townsend, 2000; Ling & Yttri, 1999). Indeed, the psychological comforts have often been pointed out by respondents to sociological enquiries (Plant, 2000).

We learn interesting things from such research about when and for what purposes people use mobile telephones in emergencies, but it still leaves the question, what kinds of data would be of use in the mobile environment?

What we might hope for could be divided into three categories: (1) Wireless networks should benefit from technologies and procedures that can give us more resilient communications systems; (2) We can better make available large amounts of high quality data, data analysis and interfaces useful in disaster situations; and (3) The ways people use mobile telephones in emergencies can be improved through training and adopting new norms of behavior that make such functions and facilities more effective. The first of these will be addressed in other works currently under development (Columbia University Center for Resilient Networks); for the remainder of this chapter we will concentrate on the second with some consideration of the third.

To some degree the specific need depends on the intended audience. We could consider a structure of targeted data sets represented in appropriate forms for different users. Emergency services workers would want access to both detailed forms of static information, such as building plans, and dynamic data such as the status of fires, the locations of people, the endangered stability of structures, and so on. People caught in a forest, or a large commercial site such as an oil field, a chemical plant or refinery would appreciate instant information about escape routes or the means to assist endangered people.

In addition to the provision of and access to data, there needs to be considerable insight into how it can be used. For trained emergency services workers, this is less of an issue because they can be drilled in protocols and procedures, but for the general public it will be important to have an understanding of how emergency workers would use the available information. Similarly, emergency workers would need to know what specific information (about for example escape routes) is currently being supplied to those trapped so that they could anticipate how they might use that information. We can divide the kinds of data that might be provided into two categories, static and dynamic. The static would consist of data about fixed facilities or accepted procedures (such as evacuation plans) that can be accumulated on a regular basis. Dynamic data and information would include real-time feedback from sensors and other automatic sources, plus the manual feed of data to targeted users about changing situations.

Static Content:

- Building plans: The landlords of new skyscrapers have access to relatively high quality graphically appropriate data about the exact plan of the building and utilities. This data, in graphical form, could be made available in a form appropriate for mobile phones and show floor plans, office lay-outs, escape routes, and other basic building information.
- Information about planned escape routes, emergency procedures, and sc forth, including contingency plans on alternative escape routes and emer

gency access that have been agreed by building operators and managers. This could be extended to evacuation instructions on a larger scale as would be necessary to evacuate a factory site, a forest fire zone, or other endangered area.
- Technical data about building structures (materials, utilities, back-up provisions, etc.): Data about the structural characteristics of buildings, including simplified assessments of their ability to withstand fires of certain levels of intensity or in specific locations of the building. Similarly, the location of flammable or otherwise dangerous materials or structures, such as generators and fuel tanks, could be highlighted when relevant.
- Neighborhood layouts: Maps of escape routes so that plans could be generated and disseminated for areas around endangered areas. This could be extended to include instructions for urban evacuation routes and procedures.

Dynamic, Pushed Content:

- Automatically updated data from sensors (heat, smoke, water, movement, radiation, etc.) can now be offered, especially as the development of ad-hoc networks of sensors can provide direct communication links to local mobile phones.
- Data provided by personal mobile devices (location, damage, etc.) can be shared on networks of nearby phones.
- Short-term instructions and corrections to or status reports about static data, such as recommended alternative escape routes. This would be information constantly made available by emergency workers or other responsible persons.
- Reports on the status of individuals or small groups: Sensors to detect the location of people and perhaps with further capabilities that include their movements and even their state of distress.

These data and information sources need to be authenticated and managed to ensure that materials are relevant and updated. This is expensive. To ensure compliance, some providers of information will need to be guided by statutory obligations. Others will need to be guided by codes of conduct.

Further opportunities abound in relation to the content potentially carried by mobile emergency systems. Some are obvious and prototypes have been built or are being considered. These include instructions and physical directions on how to escape or avoid danger. Others include ad hoc networks of sensors that feed data to mobile devices. These might include mechanisms for monitoring heat, motion, or water. They might eventually be able to include structural stability data such that impending danger from a collapsing building is measured and communicated. Advanced applications of GIS, especially when integrated into building layouts (as are currently used by some utility companies and occasionally by fire departments) offer other possibilities for content.

A considerable amount of this sort of data is already potentially available, but it is of highly variable quality, and perhaps even more troubling it is often regarded as commercially sensitive or too expensive to restructure for use in emergency work. Extensive prototypes and some commercial applications in Japan's DoCoMo i-Mode show how the use of graphic maps and games can present data, including geographical position, in an easily usable form. Local WiFi capabilities show how inexpensive transmitters can be used to provide high bandwidth access to nearby services and customized information. New models of mobile telephones, personal organizers, and other communication devices show improvements in interfaces and especially in the quality of screens. Similarly, microphone sensitivity and speaker power are already such that the telephones can be used more easily in times of stress than previously.

However, what is still absent is any indication of how large scale, standardized data can be collected and made selectively available on need. The criteria of need might be easily determined. Location would be an overriding priority, with emergency services and key government and management personnel given immediate access. But there should also be ways of determining how others, such as otherwise unidentifiable friends and relatives might be included in the prioritization. Hierarchies of material could also be specified, such that full data would be made available to people in the midst of a disaster, including building and street layouts, status reports, and so on, but only communication connections, without access to large amounts of what might be sensitive data, would be available off-site. Although we will not consider interfaces here, emergency situations have demanding requirements, especially when visibility is poor, hands are busy, or stress and distractions create severe psychological strain.

Arguably the most critical matters are not the quantity or even the absolute quality of the data but rather the utility of its presentation, the manner in which information is regarded as critical when emergency activities are carried out. The norms which are necessary to guide this change will come about slowly. But those who have personally been touched by a disaster in which they could conceive of the value of better communication will form a large group of early adopters. Norms are changing very quickly as functionality and fashion coincide to make well known certain uses.

In addition to making use of generally supplied emergency information, we might expect individuals to customize their mobile communications software for their own needs. This would be similar in practice to coding in personal addresses and telephone numbers, constructing chat group protocols, and other simple software tasks. It could become common practice, perhaps even a service offered by mobile telephone service providers, to code supplementary emergency call numbers into the telephone that might be activated in an appropriate manner. Perhaps that same function could automatically connect the caller with a central emergency information provider. That information could be accessed through a menu to bring the person to more personalized, more localized, or detailed data.

The acquisition of data will be costly, but even more costly are the management and update requirements. Further costs will be incurred if standby staff is required for emergency live feeds, new analyses, and labor intensive activities. The best way to mitigate these expenses will be to embed the emergency services with other routine activities of the communications function.

Who would be made responsible for the acquisition of data, how could it be monitored, checked for quality, and procedures standardized and enforced? Furthermore, who would pay for such services and the effort to produce initial data, and how would they be charged? One model might be a requirement on franchised telephone companies to pay for the service, through charges on customers.

A similar set of questions can be raised in relation to the quality of information supplied and the responsibility to keep it updated and accurate. Given the seemingly poor performance of data handling for infrequent, non-commercial uses (such as the erroneous, out of date spatial use data that led to the American bombing of the Chinese Embassy in Belgrade in 1999), we must be very cautious about the accuracy and timeliness of critical data. Some of these kinds of problems will surely become less significant as homeland security and earthquake disaster awareness rises.

Many of the suggestions offered so far would be unacceptable to some interests. Some of them violate common norms of what is inappropriate to share, such as office layouts. Some go still further and challenge data privacy. Even beyond these issues, commercially sensitive matters such as office facilities and the presence of and location of special equipment might militate against getting some of this information. One might hope also that a proper balance could be struck between intrusiveness and security on the one hand, and the value of the data for its utility in times of emergency.

3 Conclusion

The dreams of extensive networks of third-generation mobile telephones, or some more advanced, Internet protocol standard mobile communication devices have been directed to entertainment and commercial applications. While further imagination will bring about other uses perhaps more appropriate to our personal situations and needs, thinking about the functionality and content of mobile communications devices in times of emergency gives us a higher goal, one of importance to our lives and even large scale social interests. The opportunities here are extensive and deep reaching. They imply a new focus on the individual's needs, from escape and assistance to providing the means to communicate under conditions of destruction, stress, and surrounding chaos. But they also demand much imaginative thinking about the kinds of information that emergency workers can use and provide, how to manage the production and maintenance of that content, and perhaps most importantly, revisiting the most basic challenge of ensuring that these capabilities are appreciated and used.

References

Anderson, P. S. & Gow, G. A. (2000). *Commercial Mobile Telephone Services and the Canadian Emergency Management Community: Prospects and challenges for the coming decade.* Office of Critical Infrastructure Protection and Emergency Preparedness, Canadian Ministry of Public Works and Government Service.

Chapman, S. & Schofield, W. N. (1998). *Lifesavers and cellular Samaritans: emergency use of cellular (mobile) phones in Australia.* Typescript 31 January 1998.

FCC. (2001). *Docket No. 94-102 and WTB/Policy fact-sheet.* January 2001.

Geser, H. (2002). *Towards a sociological theory of the mobile phone.* University of Zurich, Typescript August 2002.

IBM. (2002). *Government mobile and wireless scenario.* IBM Lotus Software—Mobile & Wireless Solutions Support, Mobile Notes.

Kapsales, P. (2004). Wireless messaging for homeland security. Using narrowband PCS for improved communication during emergencies. *Journal of Homeland Security* [March]. Retrieved from the World Wide Web: http://www.homelandsecurity.org/journal/articles/Kapsales.html.

Ling, R. & Yttri, B. (1999). *Nobody sits at home and waits for the telephone to ring: Micro and hyper-coordination through the use of the mobile telephone.* Telnor forskning og Utvikling, FoU Rapport 30/99.

Noam, E. (2001, September 24). New Economy; Straining communications systems. *The New York Times.*

Noam, E. & Harumasa, S. (1995). Kobe's lesson: Dial 711 for 'open' emergency communications. *Telecommunications Policy, 19*(8), 595-598.

Noll, M. (2001, October 17). *Closing remarks.* Columbia Institute for Tele-Information, "Disaster recovery workshop; communications and information systems in the financial industry: Lessons from the city of London and world trade center attacks".

Palen, L., Saltzman, M., & Youngs, E. (2000). Going wireless: behavior & practice of new mobile phone users. CSCW '00, December 2–6, 2000, Philadelphia, PA.

Pershau, S. (2003). Multimedia applications. *Presentation to ITU workshop on Telecommunications for Disaster Relief,* 7-19 February. Retrieved from the World Wide Web: http://www.itu.int/itudoc/itu-t/workshop/ets/s7p1.html.

Plant, S. (2000). *On the Mobile. The effects of mobile telephones on social and individual life.* London: Motorola. Retrieved from the World Wide Web: http://www.motorola.com/mot/documents/0,1028,333,00.pdf.

Stephenson, R. & Anderson, P. S. (1997). Disasters and the Information Technology Revolution. *Disasters, 21*(4), 305–334.

Townsend, A. M. (2000). "Life in the realtime city: mobile telephones and urban metabolism". *Journal of Urban Technology, 2*(7), 85–104.

17

Access of Content to Mobile Wireless: Opening the "Walled Airwave"

Eli M. Noam

1 Introduction

As wireless communications progress, they encounter as well as create new barriers to openness. This chapter will discuss the problems of access by content providers and portals to wireless networks.

Openness is more than competition. Competition means the ability of companies to contest each other and to seek customers' business. This can result in efficiency and enhanced consumer welfare. But it can also result in a competition among bundled product packages instead of competition on a product-by-product basis. Openness, on the other hand, means the ability of competitors to access consumers directly rather through their own rivals. This is particularly an issue in network industries, and has been a constant theme of regulatory battles for more than a century. In telecommunications, product and service markets were closed to competitors for a long time. For example, rival equipment makers existed domestically and internationally but could not reach customers of AT&T's network.

Telecom networks were opened first to customer equipment. Then, openness reached long distance, international service, network equipment, and local telecommunications. It has now been partly extended to Internet service over cable TV networks. But it has not yet reached wireless communications. Here, competition has been fostered but not openness. To the contrary, most trends of wireless policy have been in the opposite direction. This has direct implications for the access of content.

2 The Problem of Vertical Integration

Mobile communications are becoming the front-line communications device for most people. After the attacks on American cities on September 11, 2001, it was used from airplanes, from under the rubble, and as a substitute for congested landlines. Wireless is moving into Internet access, transactions, and media content.

The major problem with the emerging wireless environment is that it is vertically integrated in ways that have become unthinkable for other media. Could one imagine a telephone carrier that can limit user access only to its own Internet portal that can select the accessible websites that can control the type of telephone equipment its users are attaching, and the software that these users are downloading? These limitations have not been particularly noticeable in the past, where cell phones could be thought of as some kind of advanced cordless phone for the car. But cell phones are now becoming much more than that, more like computer and media terminals on the go, and for more people. (A similar type of emerging issue is the access to interactive digital TV.)

The main characteristic of the wireless business is that the customer is a contractual subscriber who is served vertically by a wireless carrier that provides a full set of services. The basic components of a wireless operation are graphed in the following:

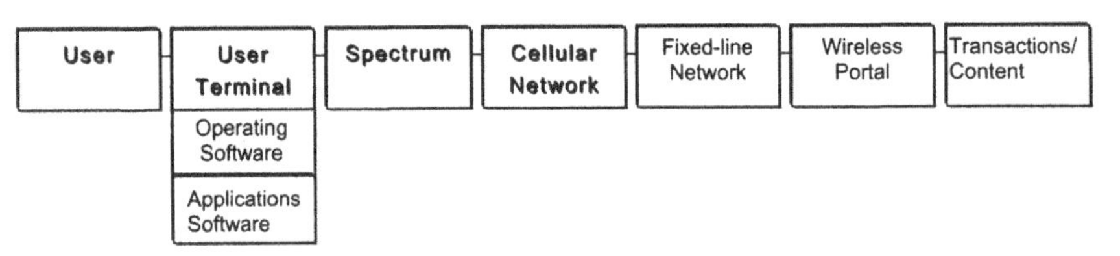

Figure 1: Components of wireless carrier's services

The user reaches his carrier via his terminal (handset) and the carrier's allocated frequencies. The call is then routed via a fixed network to the recipient. More recently, this has been extended to route Internet-style communications to a wireless portal that links into transactions and content.

The key component to this system is the spectrum allocation. It enables the carrier to control downstream the terminal equipment and access of a subscriber, and leverage this position of "owning the customer" upstream to the other steps of this chain.

In consequence, we are quite used to the notion that the carrier:

- Controls the access to a wireless portal, its content and features, of the providers linked by that portal, and of the placement of these links

- Selects, markets, and approves the customer handsets and connects it to its network
- Provides, selects, and adopts many of the features, capabilities, and content resident on the handset
- Provides software-defined functionalities on the network
- Selects and approves services resident on the network and provided by itself or by third parties
- Operates the wireless portion of the communications path
- Operates or provides the local fixed line distribution
- Operates or selects the long distance and international carrier
- Selects, for areas in which it does not provide service, a partner mobile carrier that services the subscriber, at rates negotiated and billed by itself

There is nothing wrong with a carrier offering all of these components in a bundled fashion. However, one can readily recognize good old friends, issues that have bedeviled fixed line telephony and cable television:

- Selectivity over content, which would be particularly troubling as the wireless medium becomes a mass medium with audio, text, and maybe video
- The reduction or lack of customer choice in applications and content inherent in a vertical integration with no or limited alternatives
- The reduction in innovation of service provision due to the closed nature of the applications and software that can be offered by third parties
- The absence of choice for customers to use, where more advantageous, alternative wireless arrangements are possible such as wireless LANs, other carriers for roaming, or stronger signals of another carrier
- Market power with respect to vendors of m-commerce, and requirements on such vendors to become business partners
- Restrictiveness in the inter-carrier transfer of instant messaging.

These problems will now be analyzed in greater detail.

3 The Problems of Wireless Non-Openness

3.1 Closed Portals Reduce User Choice

Under the presently evolving system, users reach a wireless portal, from where they can be connected to a variety of other sites. The selection and placement of these links, however, is under the control of the carrier. Other portals might be accessed, but that requires additional clicks. This situation is very similar to the one discussed for cable television's access to portals other than those of the cable company or its partner. These issues, subsumed under the term of the "walled garden", are well known and require no recapitulation. Virtually the same arguments on both sides apply also to cell phone access to portals, and through them, to the broader Internet. It should be noted, however, anticipating the conclusion of this paper, that they are much easier to resolve for the wireless medium.

3.2 Transactions and Content are Limited

The wireless carrier's portal is not a common carrier. Hence, the selection of websites, e-vendors, and content providers is entirely that of the carrier. Its selection would be based on its own economic, cultural, and political considerations. Being a selector, it would also incur some legal liability, which would further increase caution.

3.3 The Usefulness of User Equipment is Limited by Closed Operating Systems Software

Beyond the question of whether multi-services equipment can be licensed and connected is the question of control over the nature of the terminals themselves. As handsets become smarter, they begin to resemble small computers. To function, they incorporate operating system software.

As wireless networks begin to offer increasingly higher-level services, the question of who may load what applications onto a handset, and what network-based service interfaces these applications may access becomes important. Is a user restricted to only the applications that are offered by his primary service provider, or may he load other applications? Furthermore, can these applications have full access to the functions of the network and the handset?

3.4 Reduction of Choice Among Cellular Service Providers

Currently, cell phone users enter into a service agreement with a single carrier. That carrier accepts all of their calls or reaches them in the case of incoming calls. Where the user is outside the service territory of the carrier, the user is serviced by another carrier in a "roaming" arrangement. The roaming-partner carrier is selected by the primary carrier in a com mercial agreement (a "preferred" roaming arrangement). The call could also be picked up whichever carrier is around (a "general" roaming partner, typically a set of carriers, with prices set industry-wide). The third type of arrangement is based on signal strength, where the roaming goes to the strongest signal in that area, unless there is a primary or general roaming agreement, which would override. Whatever the arrangement, the user has no choice in the matter, in contrast to the arrangement in GSM countries, where a caller can select the roaming carrier and override its primary carrier's choice.

In the U.S., this choice is further limited by the different wireless protocols used by carriers. A user of a carrier operating on the CDMA standard cannot roam, in technical terms, on a TDMA or GSM carrier. A limited number of handsets can use both TDMA and GSM since they are related. But on the whole, the ability to switch to a carrier using another standard is minimal. In contrast, in GSM countries users can easily take their handset to any other carrier.

Furthermore, it is impossible to subscribe to more than one carrier using a single handset. For example, if a user spent much of his time in both New York and Atlanta, and no company serviced both cities, he might want to subscribe to companies in both cities rather than pay expensive roaming charges. However, there is presently no practical possibility to switch between two carriers. In theory, something exists called "dual NAM" that would permit dual-carrier subscriptions. In practice, however, phone inquiries to several major carriers did not reveal the availability of such arrangements. This contrast with the situation prevailing in GSM countries, where user can have the "SIM" cards of several carriers and inserts one of them into the handset when she wishes to use that carrier.

Also in theory, a reseller or reseller group could resell the services of more than one carrier or service type. This assumes that permission would be granted by the carriers whose service is being resold, which is not likely if they refused to permit such choice for their direct customers.

This lack of choice has real implications. Roaming calls are quite expensive, and are not part of the subscriber's "bucket" of minutes. They are a major moneymaker for carriers.

The main problem here is not technology but resistance to competition. Once a user can switch freely among carriers, where will it end? A user might regularly drive through some areas where the signal of his primary carrier is missing, and then select another carrier that performs better. Next, a user might switch to a carrier who offers her the lowest rate during that time period. Soon, the user would be able to engage in "least cost routing", LCR, as in "always best connection" (ABC). This means that there might be automated competition for every call, as opposed to the present system of competition for the subscription.

3.5 Absence of Choice Among Different Wireless Services

In the past, cellular phone service constituted an end-to-end service, separate from those of others. However, other wireless services are also being offered. Paging has long been a widespread service, and smart paging via narrowband PCS has gained increasing popularity. An example is the BlackBerry pager for always-on email. Some such services are being offered on cell phone terminals, but only using its cell phone frequencies, as opposed to being able to switch to the service provided by another paging company. Furthermore, a cell phone terminal could conceivably be used as a terminal for a cordless phone at home or at the office, directly without going through the wireless network. Similarly, it could be used as a "walkie-talkie" between several other cell phones in a neighborhood, again without going through the actual network. (This is a popular feature provided by Nextel for its own subscribers). It could be a terminal to the type of data services pioneered by Ricochet. The cell phone terminal could also bypass the wireless network through wireless local area networks (WLANs). Or, the cell phone terminal could be used as a radio receiver for broadcast programs, a scanner for police frequencies, an advanced pager, a ham radio, marine radio, and so forth. It might be used in a peer-to-peer fashion, by passing carriers altogether. It is time to think of what we now call the cell phone handset as a future general multipurpose wireless terminal. Not as an end point of a specific wireless network but as the starting point to use applications, using whichever wireless system fits best.

Such multipurpose terminals would be a threat to most cellular carriers. To see that, let us consider the case of public and private Wireless LANs (often called WiFi networks) that are emerging as so called "hot-spots" on college campuses, airports, office parks, coffee house chains, apartment house complexes, and planes and trains. These networks, operating on unlicensed spectrum, already reach wireless speeds of up to 45 Mbps two-way communications, and can service, in principle, any type of wireless device, whether laptops, PDAs, pagers, or mobile phones. They follow the 802.11b standard advanced by Apple, or the Bluetooth standard (whose range is more limited), or the emerging HiperLan2 standard.

These WLAN's are expanding into short-range "home networks" as well as wide area wireless Internet service providers (WISPs). These advantages are cheap and easy installation, use of unlicensed spectrum (i.e., without the cost and delay of a licensing process), and flexibility to change to the next level of technology. Entry barriers are low and could include hotels, colleges, airports, shopping malls, and so on. Disadvantages of WLANs are lower security, the need to coordinate billing and roaming, and the low staying power of new entrants. (Two early entrants – MobilStar and Ricochet, have gone out of business. The latter aimed at a national coverage.)

At the same time, the cellular carriers' 3G plans are also being contested from below, from upgrades in the second-generation technology known as GPRS, EDGE, and others, generally called "2.5 G." These technologies raise the data rate for mobile operations to speeds not greatly lower than those realistically expected by the third generation UMTS.

Hence 3G operators are in a bind: They often paid high prices for their new licenses, their average revenue per user (ARPU) is lower than in the past due to competition, and their new data business might be crippled by a combination of WLANs and 2.5 G. Such a combination with the flexibility of software-defined radio technology and unlicensed spectrum, might give use to the next generation of wireless—"4G"—that would be characterized not so much by superior technology but by more flexible one.

3.6 Control Over the Approval of Handsets Reduces Innovation and Choice

The carrier's business calculus on what equipment to approve is based on a variety of factors. Since in the U.S., in contrast to Europe or Japan, the carrier rather than the consumer buy most handsets, low cost is a major factor, as would be serviceability, ability to maintain a limited inventory, and independence from a single source. In addition to reducing the choice available to users, this system also makes manufacturers somewhat dependent on large carriers. The handset makers also tend to be major suppliers of network equipment. They would not lightly put used equipment into the marketplace that would be disfavored by the carriers as threatening their basic business by facilitating access to services such as WLAN that compete with the business of their best customers.

The absence of openness resembles the "walled garden" arrangements of some Internet portals provided by cable companies. Correspondingly, we can call this arrangement the "walled airwave" system.

4 Implications for Public Policy

The previous section has identified the potential for real problems. But the recognition of such issues does not mean that regulatory approaches are needed. A vigorous competition among mobile carriers could overcome most issues and generate unbundling through market forces. At the same time, the ability to exercise market power with respect to mobile commerce providers or wireless LANs might be common to all mobile providers and more profitable than a more open system. In such a case, market forces might not lead to unbundling.

The knee-jerk response to the problems identified in this paper is that competition will take care of it. But suppose that carriers would be consistently worse off by offering consumers the choice of moving easily around to other carriers or service providers. Such competition would reduce prices and profitability. It would, on the other hand, grow the market. But it is quite likely that each carrier would be better off servicing a less competitive slice of a smaller market, rather than engaging in greater competition in a larger market.

It is not clear why a carrier A would be the first to offer such choice to its customers. After all, it would provide an exit to its own customers, without a potential compensating gain from the customers of the other carriers B and C. The main reason would be to hope for enough users of B and C to switch their subscriptions to A in order to have the choice of not using A. This can hardly be a strong selling point. Furthermore, any choice requires the consent and cooperation of B and C, which might not be forthcoming once they realize that they are opening the door to a mutually destabilizing competition. They will be concerned with reputation effects if they are blamed in users' mind with poor performance caused by an element not under their direct control. And they might be able to use bundling as a way to price discriminate, as George Stigler has pointed out in a different context. The likelihood of oligopolistic behavior within a small group of carriers is high. As the number of competitors shrinks, each has less to gain and more to lose by maverick behavior. It is also an inhibitor for any software developer to take initiatives for new applications if the market is largely closed, and this further reduces the attractiveness of any non-conforming behavior by a carrier.

Where market forces do not work, would regulation? Let us look at several potential points of intervention and evaluate their need.

A schematic view of an unbundled wireless network environment is provided in Figure 2. It shows, at each stage of the chain of wireless provision, alternative providers. We conclude that only one of simple policy—the openness of the terminal equipment to access multiple providers of wireless services and providers—is critical. A subsidiary second opening—spectrum—supports such policy.

4.1 The Separation of the User Equipment (UE) from the Carrier

Such a policy would amount to a *Carterfone* policy for users' wireless equipment. Following that decision in 1966, the FCC permitted users to attach equipment chosen by themselves to the telecom network. While the carrier could still offer and market its preferred equipment, it could not exclude other equipment, as long as it conforms to certain technical specifications pertaining to the RF transceiving function and non-discriminatory industry specifications for air interfaces standards. These specifications could not close equipment third-party applications or access to other network protocols offered by other types of providers, as long as it conforms to the FCC's software defined radio rules.

While a fully bundled service could be offered by a carrier as before, the carrier could not prevent a user from selecting, for any given call, another wireless service provider or using the equipment for other communications purposes.

The significance of such arrangement is that equipment will be offered by the market that adds features, and, more importantly, permits a user to select service providers depending on circumstances. For example, a user in a shopping mall, campus, office building, or airport could connect to a wireless LAN. A user encountering a circuit busy could switch to another carrier. A user seeking to receive synchronous music, radio style, could do so by accessing a specialized broadcaster.

This choice would reduce the need for most other access requirements, since the user would not be tied to a single carrier with significant costs of

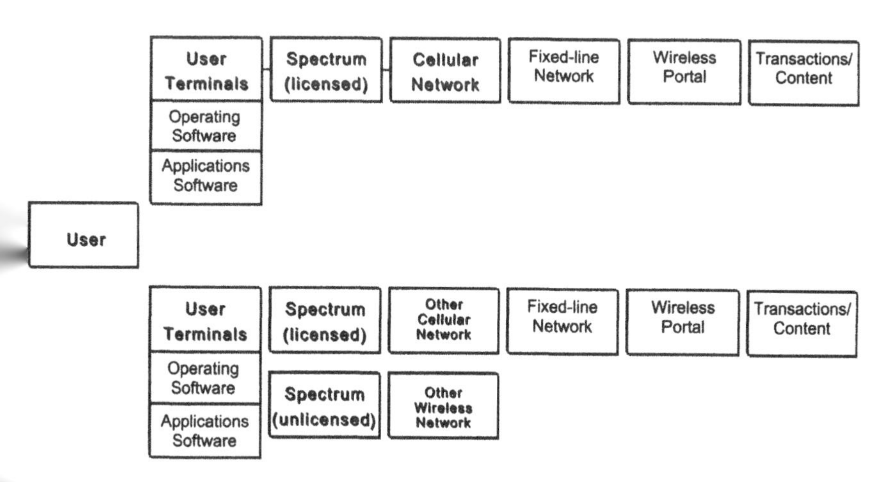

Figure 2: Schematic view of an unbundled wireless network environment

switching to another. This is partly embodied in the GSM standard which provides some user selectivity over carriers, although the approval of such alternatives remains with the primarily carrier, which also handles the billing.

This approach would be similar to that adopted by the FCC for CPE following the *Carterfone* decision in 1968. It followed Cassandra warnings of impending network chaos, but has worked spectacularly well.

4.2 Access to Unlicensed Spectrum

The key source of leverage for carriers is the high entry barrier for new and future entrants in service provision, due to the spectrum auctioning system with its advance payment feature. Given the difficulty in freeing additional spectrum and the high cost of acquiring it, it seems unlikely that there would be new entrants emerging to challenge the reduced group of carriers. Therefore, government should additionally provide adequate spectrum on a license-free basis, with users and service providers paying for usage rather than for ownership, in the way that automobiles pay for the use of highways. This has been developed in detail by the author in other papers.[1] Once such spectrum is available, and once users' terminals can access service providers such as WLANs operating on such spectrum, users will not be constrained by the limited choice of maybe four cellular carriers that could still collectively be restrictive.

4.3 Access to Alternative Wireless Portals

The third access issue is that to the wireless portal. The issues here are similar to those discussed for the cable industry. The similar arrangement would mean that the wireless carrier would let the user pre-select its primary portal, or that several such portals would be accessible at no extra effort, or that the two upper layers of the carrier portal would be open to third parties. This approach would mirror the open access of the Internet, and the approaches now being applied to Time Warner and considered by the FCC in its proceedings.

Content openness may be the easiest type of openness to consider since it is essentially a browser level openness. The question can be reduced to whether the user can enter an arbitrary URL to a network portal to access content (independent of any business deal between the wireless provider and particular content providers) and whether browser plug-ins can be created and downloaded to render the resulting content if required. This issue is analogous to the walled garden discussions that have occurred in the wired Internet.

This problem would largely go away if the users could also access, through their handsets linked to other carriers and wireless providers, other portals and websites.

4.4 Openness of the Carrier's Network

The fourth element of openness relates to services offered by third parties and requiring presence in the wireless network. The options are either to keep wireless networks closed to third parties, or total openness, resembling a common carrier access for third party software applications, or a type of equipment collocation that exists in telecommunications. Here, too, the ability to access alternative wireless carriers through flexible handsets would be enough to deal with this issue.

5 Conclusion

The focus of government's policy has been to provide wireless carriers with choice—in the utilization of the licensed frequency, in the technical specifications of it service, in its pricing, and so on. There has been no similar orientation towards choice of the users, for content and transactions. The implicit notion was that by establishing rival carriers users will be well served. That strategy certainly goes a long way. But carriers are likely to resist offering consumers the choice of moving easily around to other carriers or service providers.

The conclusion of the analysis is that the key point of openness, and arguably the only one needed, is that of *openness of user equipment*. With this openness achieved, the user would have alternative avenues to spectrum, content, portals, applications, software, and so forth. A secondary policy would be to assure alternative wireless pathways such as WLANs by providing an adequate amount of unlicensed spectrum.

Why is all this important? The overall goal of the openness approach described above is to establish for the wireless and wireless content environment the same dynamism shown in the Internet with its open access terminals encouraging hardware and software innovation and applications. Right now cellular telephony is a dynamic sector, mostly based on the growth of penetration. Soon, however, this growth will plateau as universal wireless connectivity is being approached. At that point, we need the impetus for further innovation that a more open system provides. For the carriers, the overall positive impact in terms of traffic generation may well outweigh some loss of control. For users, service providers, and technology developers, the advantages of openness might be significant.

Communications policy has fared best when it puts its faith in the dynamism of the periphery of the network, instead of seeking to strengthen the ability of the network core to dominate. Wireless is no exception. And the mediocre results of policies focusing on the core, in contrast to those for other parts of the communications environment, suggest that a reorientation is in order. With it we can leapfrog the "3G" model with its carrier-orientation to a "4G" model patterned after the Internet, and overcome the "walled airwave" problem.

Acknowledgments

I am grateful for the help and comments received by the following people: James Alleman, Bob Atkinson, Ron Barnes, Brian Bebchick, Kenneth R. Carter, Kathryn Condello, Terry Hsiao, John Lee, Don Nichols, Michael Noll, Michael Marcus, Bertil Thorngren, John Williams, and to Charlie Firestone and the Aspen Institute's Regulatory Policy Meeting, especially Kevin Kahn and Robert Pepper. Views expressed here are entirely my own.

Endnotes

[1] See Noam, Eli M. (1998). Spectrum auctions: Yesterday's heresy, today's orthodoxy, tomorrow's anachronism: Taking the next step to open spectrum access. *The Journal of Law & Economics*, XLI, Part 2, 765-790.

References

Bell, T. W. & Singleton, S. (1998). *Regulator's Revenge: The Future of Telecommunications Deregulation.* Cato Institute.

Carterfone, 13 FCC 2d 420 (1968).

Hazlett, T. W. (2001). The wireless craze, the unlimited bandwidth myth, the spectrum auction faux pas, and the punchline to Ronald Coase's 'big joke'—an essay on airwave allocation policy. *AEI-Brookings Joint Center Working Paper* (01-2).

Lehr, W. & McKnight, L. (2005): Wireless Internet access: 3G vs. Wifi? In ECC (Ed.), *E-merging media. Communication and the media economy of the future* (pp 165-180). Berlin: Springer.

Lemley, M. A. & Lessig, L. (2001). The End of End-to-End: Preserving the Architecture of the Internet in the Broadband era. 48 UCLA L. *Review* 925.

Petty, R. D. (2003). Wireless Advertising Messaging: Legal analysis and the public Policy Issues. *Journal of Public Policy & Marketing, 22*(1).

Pitkänen, O., Soininen, A., Laaksonen, P. & Hurmelinna, P. (2002). Legal constraints in readiness to exploit mobile technologies in B2B. *International Workshop on Wireless Strategy in the Enterprise*, Berkeley, CA.

18
Mobile Mass Media: A New Age for Consumers, Business, and Society?

Jo Groebel

1 Mobility Revisited

Mobility—a magic word in the 1950s. Futurologists dreamed that the "PH", the personal helicopter, would become reality towards the end of the millennium. They fantasized about a limitless universe that extends the known world—using new motor power—into new regions upwards in space and downwards in the oceans. While some of these visions have become reality, the former "future", that is the present now, takes mobility for granted but certainly does not define society primarily in terms of an age of physical movement. And reality has taken on a different direction.

In the 1990s, the new magic word was information. And when the e-business dream unfolded, futurologists, who by now called themselves "entrepreneurs", projected their ideas into a bright and shiny society based on knowledge as the new driving force. Soon, the explosion of wireless communication technology also initiated a revival of "mobility". It was added to the concept of the information age and fantasies revolved again around a 24-hour world of unlimited business, work, and fun. In Europe, fuelled by these hopes, telecom operators invested billions of dollars to secure their position in the future prosperous mobile market: For example, UMTS-frequencies, were auctioned off in Germany alone for approximately $50 billion.

However, the digital dream collapsed in the early 21st-century. And yet, mobile devices had become the common equipment of a majority of consumers in large parts of the world. New unpredicted "bottom-up" applications like SMS or MMS turned out to be very successful and demonstrated the dynamics of the market.

In such a situation, once more the rule may apply that, in technological development, short-term changes are over-estimated but long-term changes are under-estimated. While the hope of an immediate digital economic revolution was not met, the evolution into a different world driven by, among others, mobile technology will probably move on. Too broad are the advantages of wireless communication beyond voice-to-voice applications, too affordable the devices not to assume major changes in business and social paradigms based on the "new mobility".

2 The Digital Emancipation from Time and Space: 21st-Century Mass-Media Content

2.1 Mass-Media Socialization in a Flexible Environment

Mass-media content traditionally demands users to make time and space decisions. In order to obtain a newspaper or a book consumers have to go to a newsstand or to a store. A newly released movie is only available by going to a movie theater. TV shows are usually shown only once or at best a few times. Until recently, audio-media like radio and cassette or CD were the most mobile electronic content providers. With transistor technology, the radio became mobile in the 1960s, the walkman became a household name in the 1970s, the watchman did not make it in the 1980s. Again, it was an audio-medium, the disc-man, which won in the 1990s. Now, it is the MP3 player.

The reason for the success of these devices, apart from the device size, is their behavioural function: Music and sound are potentially low-attention media; they can be used as parallel-media while being involved with other activities. Or they answer mood-management needs when they reinforce an already existing emotion in a given situation, for example, a romantic date. Thus, they are a complement or an accompaniment to another focus. Portable games, from the early Nintendo-offers to the sophisticated game devices of the early 21st-century, however, are high-attention media. These observations already hint at the important role of the situation of media use. Despite the higher convenience of nearly any mobile device, rituals, practicabilities, and behaviour patterns still determine what kind of local or flexible media content is preferred.

However, times are changing. *Media socialization* along with availability, convenience, and price affects the direction of future individual behaviour. The generation that was brought up with the laptop and the wireless phone expects a broader spectrum of mobile content. They are increasingly used to services which meet the "5-i"-criteria: immediacy, internationality, integration (of media), independence (of time and space), and interactivity. To give an example: SMS allows an even more spontaneous distribution of information than audio-telecommunication; it crosses the borders without technological or regulatory constraints; it may be the basis for multimedia messaging; it can be sent and received with any preferred delay; and of course it is interactive. SMS may pave the way for other forms of dynamic communication. The digital environment has become potentially independent of time and space: text devices are mobile. Laptops, PDAs, new-generation mobile phones not only offer opportunities for individual communication, they also process downloaded mass-media text content. The 40 printed volumes of the "Encyclopaedia Britannica" are now a CD-ROM or an Internet-download away. The reason why e-books have not quite made it may be due to the fact that they are not really needed: they could not compete with an already existing mobile medium, the pocket-book. The specific haptic qualities of bound paper also still provide a major ritual advantage.

Nevertheless, modern media socialization means that the early 21st-century witnesses a quiet revolution of time-and-space independent media (Noam, Groebel & Gerbarg, 2003). The digital cinema of the future is fed by satellite-distributed movies. If it were not for the theater as a location of high-quality sensual and social attraction or for the financial need to keep certain value exploitation chains, it could be skipped altogether in favor of a direct link to the home. This would certainly be facilitated by the success of the home theater with LCD or large plasma screens and Dolby, THX, or new forms of surround sound. For movies, the time step until they reach homes or people on the move has melted down. Discounting piracy, the time-span between first release, and following video-on-demand, DVD, pay-TV, and free-TV distribution has remarkably shortened. Potentially, the DVD makes a movie mobile within a few months of opening night. Yet, small devices are inferior means for movie watching. Who could imagine watching "Gone with the Wind" or "Gangs of New York" on a small mobile display? Again, there is a relation between content format, its function, and a viewer's mobility need.

Further indicators of the increasing electronic flexibility revolution are the hard-disc recorders (PVRs) for TV that challenge the fixed-schedule watching of programs, as well as the navigation based car TV-screen, and particularly MP3 players and other data-compression technologies for music. They have moved from limited-storage media like vinyl, cassette, or CD to downloaded "unlimited" hard-disc sound pools with thousands of songs on a small device. The music industry still does not have a convincing business model to respond to this move from physical to virtual space that may replace the physical CD with some form of subscription service.

2.2 From Multimedia to Polymedia

Despite the prepared grounds for a new age of electronic mobility, one major challenge to the breakthrough of all kinds of mobile content remains: the lack of systems-interoperability. The laptop and some PDAs may integrate many functions of information and communication. Yet, the small device dream-machine which is the epitomy of digital convergence does not see the light of the day because of a lack of technical commonality, collaboration, common platforms, copyrights, and competition. UMTS could be a potential basis for the small intelligent device linked with different peripheries across time and space: the hardware would carry the mobile digital center to link any service subscription to movies, music or other mass-media with the periphery devices that will be used for media consumption such as plasma-screen, HiFi, or an Internet-booth. This goes beyond the 1990s idea of one single monitor that unites all contents, known as *multimedia*. Now, we have an emerging intelligent mobile center which hooks up with the adequate local device via UMTS, W-LAN, or Bluetooth, which could be called *polymedia*.

3 Perspectives for the Mobile World: Consumer and Society

3.1 The Consumer

Media socialization is not only driven by convenience and a good financial cost-benefit-relation. They are necessary but not sufficient conditions for the adaptation of a new or recent technology. Although prospective consumer research usually can "prove" the advantages of a new product, reality often shows a random success in the market. Human behaviour is too complex for easy forecasts even with the most convincing technology.

If it were about convenience only, mobile content would make it big: it is available 24 hours, at all places, it can be customized, and it offers the 1990s' digital dream on convergence between communication, information, transaction, entertainment, and coordination. The communication part has indeed become a great success. And SMS became an unexpected hit. Yet, other parts of mobile technology remain hidden. Whether it is song-recognition software (you hum a tune into your phone and receive song title and artist immediately) or hotel booking services, these service providers are not close to a mass-audience.

Although forecasts are nearly impossible, some necessary conditions for mobile content success can be defined. These parameters are related to the fact that human behavior, despite much variation, shows some continuity across time and space. People need orientation, they seek distraction, they like to communicate with others, they appreciate doing and controlling things, they want to be thrilled. In other words, the psychological dimensions of cognition, emotion, interaction, action, and surprise are a basis for determining mobile content success.

Cognition. News and other useful information has always been regarded as a major driver for any (mobile) content considerations. Yet, surprisingly, the mobile phone has not become the center of news-gathering for users despite some interesting service-offers. Even during the September 11 shock the devices were mostly used for audio-communication rather than for mass-media texts. People rushed instead to the nearest TV-screen or radio-set, and even the Internet was dominated by personal e-mail exchange. Only when users want to access information while they are involved with other activities, for example, during an official meeting or while travelling, does the small size of the mobile unit become an advantage.

Nevertheless, mobile news and information content signifies a paradigm change: even when ritualized with the morning and evening TV-news, people used to "go to" the information. Now, customized information is coming to people regardless of the situation they are in. The stock exchange is a good example. The pre-programmed service has a perfect fit for reaching everyone at anytime and facilitates quick decision-making. For mobile content, the cost-benefit relation including the personal value of the received information is a crucial factor. For information which creates a personal advantage, the omnipresence of mobile services is of utmost importance.

Table 1: Some behavioral principles of local and mobile device based communication

Behavioral Modus	Local Device	Mobile Device
Cognition	Person-to-information	Information-to-person
Emotion	Scheduled mood management	Mood creation anytime
Interaction	Appointment centered	Spontaneity centered
Action	Specific environment based	Situation based
Surprise	Prepared	Unprepared

Therefore, any information content for mobile devices is determined by two cognitive decision-making tendencies: the personal *importance* of the received data, and/or the ability to replace existing information- and news-seeking *rituals*. It has also been argued that the small-screen display of the mobile unit would not support moving news images. However, unlike a big Hollywood production, the fast and short-term character of newscasts would suggest a mobile application even in thumbnail-format. Customized services which address more specific interests of the individual user will probably become more important, whether it in personal leisure time preferences or professional needs. As the development of new software for small target groups does not necessarily meet a balanced cost-benefit relation, the "import" of already existing services, for example, in collaboration with mass-media content providers through strategic alliances is an adequate option. With the increasing interactivity, particularly between TV programs and the consumer (call-ins, polls, tele-shopping), the door is open to not only use the mobile as a sender *to* broadcasters but also as a receiver from them.

People believe that their information seeking is solely based on rational decision-making and cool cost-benefit considerations. Yet, it is not. Only for major challenges complex cognitive processes are applied. Mostly, daily routines determine information seeking and processing. However, as mobile content is further decreasing the need to become active and to "go to" information it is an a-priori fit medium to serve the high- and low-involvement consumer.

Emotion. Even for rational applications or in business environments attractiveness and emotion play a major role in decision-making for new forms of content. Whether users opt for new technology, whether they are interested in new kinds of programs or services is not primarily determined by rational calculation. It is the expected thrill and emotional extra-value that are major factors to consumers. This starts with the design of the hardware, focuses on the fun that users may have in applying new software and in exploring its different possibilities, and extends to the emotional value of the service and the closeness to users' own interests and identity.

Apart from the relatively stable motives which drive any leisure time, e.g. hobbies, a so-called "mood management" plays a particular role. This concept assumes that media content is used primarily in order to steer or to support one's situational emotional states. If people are in a sad mood they prefer sad music if that particular mood is "liked". Even love-sickness easily falls into this category. If people want to change the mood they go for light content. A funny movie may be the answer. Research shows that this even works unconsciously. People usually do not make long-term plans for consuming music, movies or television programs, they act more or less spontaneously without applying complex decision-making processes as they would in a business environment. Usually, the mood-management media are local. With the exception of portable CD players or hard disc recorders, most leisure time and thus mood-management-media are to be found at home. However, as mood is a not a planned psychological factor it is plausible to extend the possibilities of active mood-management to any given situation. Here, mobile content is the perfect answer. Music, indeed, has been a model for this application. It can of course be extended to any other kind of distraction or mood-related service.

In this context, cognition and emotion are parallel processes: One can passionately dive into a symphony or a rock-song, say Brahms' "First" or Aimee Mann's "Lost in Planet", and at the same time consciously enjoy their musical structures. One can also be involved with work or shopping and be pleased about a background-sound. In fact, this is a separate genre: "muzak". Cows were found to produce more milk when they are treated with music that is easy to listen to. Even students can get better grades when they do their homework backed by music. Whether this is true for all genres is yet to be demonstrated. However, the logic is there: information is processed better if a mild state of emotional arousal increases attention and a positive mood. Again, it is plausible to exploit this principle in a mobile content context. Until now, the accompanying mood-management applications, for instance, for music in a shopping environment, have to assume an average mainstream taste. Linking one's own media taste with an external environment is technologically within reach: the portable earphone-device partly "encapsulates" the user within, for example, a shopping situation. Recent laser technology seems to be able to address several individuals specifically with their own sound in an identical situation without earphones. One could imagine a shop where the customers either choose their own pre-defined local sound environment or where their portable units are, via W-LAN, linked with the periphery of the local laser sound system.

The relationship between emotion and cognition can systematically be described in terms of involvement. Depending on an individual work, an individual leisure, or a communication context, three levels are to be distinguished: low, middle, and high involvement. Table 2 shows a possible classification with examples.

Low-involvement means that any content forms pure background media like an audio- (visual) wallpaper. It never actively grabs any conscious attention. Yet, it creates a certain mood. Imagine the shop or the café with their respective sounds. For leisure and work, the customized muzak example was already explained. For communication, even if it may be perceived as a "horror" scenario for many, one can imagine a customized "soundtrack" for the

Table 2: Examples of mobile mood management content depending on the situation

	Leisure	Work	Communication
Low-Involvment	Customized muzak	Customized muzak	Telephone soundtrack
Mid-Involvment	Parallel TV	Parallel Digi Games	Live report sports
High Involvment	Mobile movies	Relaxation-break TV	Export my own music

mobile phone, for instance, in a romantic conversation. Music works in the face-to-face context, it could also work on the mobile.

Mid-involvement means that attention is not permanently, maybe not even primarily, focused on the audio-visual content, yet, it can be created anytime depending on interest or the arousal value of what is offered. Television is a good example. Maybe particularly in its potential (mobile) "thumbnail" form it may partly replace the radio as a "parallel" medium with the exception of the car. When users are actively doing something else, their attention may turn to the medium, however, for the news or for a good. In the work context, psychologists even suggest that PC, laptop, or the mobile should regularly or at random offer brief mood-management distractions in order to increase the efficiency of activities through relaxation breaks. And a group get-together during an important sports event would profit from the possibility of briefly following the more suspenseful moments.

High-involvement is one of the most interesting mood-management applications. With the already described mobile society, people regularly find themselves away from home, at a hotel, an airport, or in a train. Music and the press have already made it into these environments, for example, with a printing-on-demand version of "The New York Times" in hotel lobbies. Yet, apart from the unsuccessful "watchman" of the 1980s or the portable DVD-player audiovisual content is not yet accessible in any given situation as a whole mood-management spectrum. Even with portable hard-disc devices, the choice would always have to be limited due to a lack or expense of capacity. Here, UMTS or future generation technology comes in. Linking the intelligent mobile unit with a local periphery, for instance, the plasma-screen in the hotel room would create exactly the same mood environment as in the home. For the work context, longer breaks would afford the possibility of alerting the user with his customized news or special interest TV and every music fan knows the need to enthuse one's peers about the actual favorite sound choice. Only the small mobile unit, linked with a HiFi-periphery (via W-LAN or bluetooth) offers all the fun. These are examples of how the mobile world can answer the situation based mood-management needs of the users depending on their respective involvement levels.

Interaction. Driving forces of any medium are the desire to belong to a certain group, to communicate with others, and to interact with people. Recent findings of the "World Internet Project" (Cole et al., 2003; Groebel & Gehrke, 2003; UCLA News, 2004) show that on an international scale the Internet is primarily used for communication, that is, e-mails. Mobile content, thus, which answers the particular communication needs of people is perfectly fit to be a success in the market. SMS is a good example. A particular group, namely teenagers, turned this mobile application into a success. It is practical, it is the only medium to enable communication in nearly any given situation including the classroom and of course it is a demonstration of fashionable and hip appearance.

Gender relations are, in this particular context, an important area. The first experiments to create a mutual profile recognition software for portable devices did not succeed. People would feed their personal profile and preferences into their device which then would, if a fitting partner would pass by, create a connection. However, a more advanced application could find its market. The Internet services "friendster.com" or "match.com" with their huge success show a vivid demand for connections. As meeting people is a mobile activity, it is more than plausible to apply this kind of service to the mobile content world. For business, an additional application could extend data-mining procedures which were one driving force behind the 1990's e-commerce models from a personal profile logic to a situation-typology logic.

Action. Apart from security considerations that still apply and that may prevent consumers from using electronic devices more often for business interaction (Groebel et al., 2001), any mobile service that allows a form of transaction is, in principle, plausible. Whether it is work, entertainment, finance, or managing the household from a distance, all these areas include activities which fit perfectly with any mobile service. The fact that action is potentially a 24 hour need makes the portable communication unit fit for a situation-based "management center". Tele-working was a dream idea of the 1990s. Despite some already established platforms it has not been realized to the anticipated extent. Still, many business areas would find a much more convenient working environment if the concept of mobility and situation-based action had a more consequently developed infrastructure. Thus, it makes sense to revisit the tele-working ideas and to see how they could be applied not only to the home environment but to a mobile context. If telecommunication has become a quintessential and necessary part of the business world, the idea of mobile-supported work is not far from reality. It can be improved regarding the information processing which is also a necessary part of any job, but it can also be applied to decision-making based on electronic logic. And of course the same holds true for payment and buying transactions. Two arguments may be used to explain why the idea of product displays in the streets or on TV screens and subsequent ordering via the (mobile) phone has not been as successful as forecast. It may have to do with (1) the fact that it had not yet become a mass application and (2) that the exact analysis of the psychological necessities to support such a concept had not been answered in a satisfying way. To summarize: many people would like to use mobile devices as an action and transaction steering device. Those applications will gain ground in the future.

Surprise. In the perfect consumer world, market research is supposed to lead to the successful introduction of a new product. Reality is different. Particularly with electronic devices, spontaneous and unexpected bottom-up applications often become successful. The telephone is a good example. When it was introduced in the late 19th-century, forecasts centered on the transmission of music and on limited business applications. The rest is history. More or less the same happened with SMS. Only a few would have expected it to become a major communication tool. Thus, platforms which leave enough space for spontaneous applications and which provide do-it-yourself (DIY)-tools potentially pose successful opportunities for the mobile world.

Taken all the different human behavioral factors together, the answer of mobile content success does not primarily lie in an either-or-application. It is rather the specific combination of those factors for the development of any mobile service that is essential. Business is not only about information processing, it is also about emotions. Emotion, vice versa, is also partly based on conscious decision making, partly on unconscious mood-management tendencies. All of that relates to potential action and definitely to interaction with other people. And surprise is an essential driver in all areas of life including business.

It is not possible to create a formula for mobile content success. However, at least the discussed factors should be considered. Thus, a mobile success formula would read:

$$M = C \times F \times S$$

Mobility success is a function of content, the functionality for human behavior and the respective situation a person is in.

3.2 Society

With an integrated inter-operable and saturated high-capacity mobile infrastructure, major changes could occur in the further development of society. Whether it is information-seeking, business, or leisure-time, the dominant scheme for behavior in the traditional world is still that users, citizens, or consumers have "to go to" whatever they want. Thus, part of physical mobility still means moving towards the targeted objectives. Communication has shown some of the potential changes. With the telephone and with e-mails, communication frequencies have further gone up, but this does not necessarily reduces the face-to-face contact between people. However, mobile communication itself has shown how a situation-based behavior tendency occurs in the face of a given opportunity. In the mobile world people just like to communicate to others what mood they are in, which place they are at or what they are going to do next. Thus, communication partly has become an end in itself. The step towards other kinds of behaviors outside pure communication is only a small one.

For society, these developments would basically mean a paradigm shift from the P2S (Person-to-situation) infrastructure to a *S2P* (Situation-to-person) pattern. Services and products already accompany people to a certain extent. That logic would develop further: Whatever users want in a given situation, whatever preferences they have, it could be fulfilled on-site immediately. This creates an *option-society* where the 24-hour-universal world becomes reality. The way is paved on a small scale with the "not so silent" revolution in music services (MP3, etc.) or with people slowly getting used to always being able to make their media choice whatever the content may be (hard disc player, satellites, etc.). With only a little fantasy involved, this option society sees a social and cultural convergence moving away from a phase-oriented sequential world toward a situation-oriented parallel one. Individual behavior is often influenced by locations that are connected with certain periods of life or phases, for instance, college, university, job. Often, only the need to address people together at the same place defines those phases. However, in many areas a deliberation from fixed locations via mobile information and communication technologies would, at the same time, mean a shift from this phase-determination to a situation-oriented paradigm that is based on the needs of a particular situation. For example, linking people in the mobile world with any kind of information or training in any situation could result in a meaningful convergence of job and education. Life-long learning makes sense in a world where knowledge needs to be permanently renewed in order to keep track of work demands. After basic education, more flexible forms of knowledge acquisition and management can be realized based on the mobile *S2P*-paradigm.

More specifically, it is not demographics like age or social background and thus certain work or leisure time locations that need to determine preferences and facilities of people. It is purely the (sometimes spontaneous) motive to immediately receive or apply what is wanted. However, we still think in terms of phases when it comes to education. For example, up to 70% of people, particularly those employed in SMEs, are never systematically confronted with refreshed knowledge content after having finished their basic education. This, among other factors, leads to the fact that the elderly are not systematically linked with the most actual information on a permanent basis. This supposedly makes them unsuitable for work at a later age stage. At the same time, parts of the world see a systematically increasing gap between the time percentage which people spend on the job and the whole life cycle. This creates major social and economic problems: leaving work at 60 to 65 and still being physically and cognitively fit, yet, having another 20 or more years to live without work does not fill everyone with pleasure. On top, many economies suffer from a system where the whole community has to finance the non-working population. With the increasing gap between working and non-working time, the refinancing problem increases dramatically.

Of course, it is not for the mobile world to solve these problems. Yet, the underlying situation-based information and transactions paradigm could definitely make a difference. It would assume a permanent training- and information-based infrastructure and thus challenge the traditional age-frontiers. People would be linked with educational information at any given time and space. This associates media in a positive human-development context, unlike the frequent skeptical perspective (Groebel, 2002). Thus, the mobile paradigm opens the door for a more situation-based social approach. It supports the argument that we see an evolving society that may be called *option-society:* time and space do not determine anymore what kind of behavior is preferred, in business transactions, media consumption, work, and private behavior patterns.

References

Cole, J. et al. (2003). *The World Internet Project.* UCLA: Working Paper.

Groebel, J. (2002). Media and Human Development. In: *Encyclopaedia of the Social Sciences.* Elseviers.

Groebel, J., & Gehrke, G. (Eds.). (2003). *Deutschland und die digitale Welt.* Leske & Budrich.

Groebel, J., Metze-Mangold, V., van der Peet, J., & Ward, D. (2001). *Cyber Crime Report.* Friedrich-Ebert-Stiftung.
Noam, E., Groebel, J., & Gerbarg, D. (Eds.). (2003). *Internet Television.* Lawrence Erlbaum Associates.
UCLA News. (2004). First release of findings from the UCLA World Internet Project shows significant 'Digital Gender Gap' in many countries. Retrieved from the World Wide Web: http://cop.ucla.edu/downloads/UCLA_World_Internet_Project.doc.

Index

2.5G services 6, 71, 133, 204
See also i-mode
3G 8, 21, 57, 72, 101, 115, 133, 153, 158, 201, 231
See also standards
4G 9, 137, 231

A
Always on 21, 72, 119, 124, 134, 158, 230
Applications 3, 24, 33, 45, 70, 73, 89, 98, 120, 122, 136, 142, 189, 202, 216, 227, 239
Audio 3, 38, 61, 77, 87, 122, 191, 227, 240
Automotive telematics 45

B
Brand 58, 77, 102, 105, 110,144, 157, 175
Broadcast 15, 57, 83, 123, 190, 203, 230, 244
Business model 23, 36, 45, 57, 73, 103, 118, 133, 139, 153, 187, 241

C
Communities 79, 92, 104, 196
Content providers 7, 36, 52, 57, 61, 84, 108, 133, 135, 141, 225, 240
Context
Context awareness 91
Situation-to-person paradigm 249
Convergence 33, 88, 242
Copyright 3, 107, 187, 242

D
Design strategies 69
Devices 25, 34, 51, 62, 69, 87, 97, 115, 188, 219, 239
Digital rights management 107, 192
Digitalization 25, 35
Diversification 165

E
Early adopters 123, 220
Emergency communication 124, 215
Emergency services 51, 129, 216
Exclusive rights 177, 187

G
Games 27, 36, 78, 83, 102, 121, 147, 193, 220, 240

Global media conglomerates 165, 168
Globalization 27
GPS 62, 74, 123

H
Hot Spot 11, 135, 159, 230
See also wireless local area networks

I
i-mode 22, 59, 76, 107, 118, 133, 148, 158, 220
Instant messaging (IM) 27, 66, 98, 227
Interconnection 206

L
Licensing 10, 102, 174, 187, 207
Location-based services 49, 74, 119, 129
Lock-in 157

M
Mass media 3, 47, 72, 87, 98, 153, 181, 187, 239
Micro-payment systems 110
Miniaturization 25
Mobile commerce (m-commerce) 49, 119, 122, 139, 232
Mobile Internet 22, 62, 98, 140, 174, 187
Mobile virtual network operator (MVNO) 143
Multimedia Messaging Service (MMS) 24, 40, 71, 81, 134, 239

N
Navigation 25, 52, 79, 89, 125, 241
News design 87
Networks
Adhoc networks 60, 99, 206, 219
Network formation 141
Next generation networks 203
Personal area networks 99

O
On demand 3, 21, 89, 123, 177, 193, 204, 241
Openness 210, 225
Open access 143, 189, 234
Openness of user equipment 228

P
Paid content 103, 129
Peer-to-peer 66, 97, 137, 192, 230
Personalization 139
Personal media files 100
Polymedia 242
Portal 141, 145, 179, 192, 205, 225
Premium content 118, 147
Price 49, 153
Average price per minute 24, 153, 157
Bit rate price 4
Price discrimination 197, 210
Price per minute of use (MOU) 4
Privacy 29, 74, 109, 116, 127, 221
Profitability 50, 133, 166, 202, 232
Public policies 194, 232
Policy battle 198
Public-private relationship 29

R
Real time 3, 16, 49, 62, 90, 123, 193, 218
Regulations 58, 127, 156, 168, 189, 225, 241
Revenue sharing 136, 145
 See also sharing
Ring tones 72, 101, 136, 144, 147,
Roaming 99, 128, 135, 153, 207, 228

S
Security 24, 48, 103, 121, 209, 216, 231, 247
Situated documentary 89
Spectrum
 Licensed spectrum 9, 193, 203
 Spectrum auctions 10, 234
 Unlicensed spectrum 9, 193, 202, 207, 230, 234
Sharing
 content sharing 15, 102
 file sharing 102
 revenue sharing 136, 145
Short message service (SMS) 39, 62, 73, 101, 117, 133, 159, 193, 204, 239
Society
 Mobile society 246
 Option society 249
Standards 5, 36, 53, 85, 127, 153, 188, 193, 233

U
UMTS 21, 37, 134, 231, 242
 Usability 79, 122
User-generated content 62, 100

V
Value chain 23, 41, 46, 51, 57, 201
Vertical integration 135, 226
Videophone 32, 122
Virtual storage 100

W
Walled garden 78, 129, 201, 209, 228
WiFi 13, 57, 71, 91, 115, 133, 159, 188, 201, 220
Wireless local area network (WLAN) 10, 27, 90, 115, 119, 230
 Public WLAN 135, 136
 See also Hot Spot

Y
Young people 38
 Lifestyle 29, 108, 166
 Teenagers 117, 119, 159, 247

For Product Safety Concerns and Information please contact our EU representative GPSR@taylorandfrancis.com Taylor & Francis Verlag GmbH, Kaufingerstraße 24, 80331 München, Germany

Printed and bound by CPI Group (UK) Ltd, Croydon, CR0 4YY
01/05/2025
01858438-0003